Watts Pocket Handbook

The essential guide to property and construction, as used by professionals since 1983

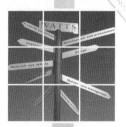

First published in 1983, the Watts Pocket Handbook is an annual publication.

The editors wish to express their indebtedness to previous editors of the handbook, who laid the foundations on which this 21st edition has been built.

The editors would also like to acknowledge with thanks the contributions of individuals in the property market and construction industry who send suggestions and comments. Such contributions are always welcome and should be sent to Samantha Rumens, Communications Executive, Watts and Partners, 1 Great Tower Street, London, EC3R 5AA.

No guarantee is given by Watts and Partners or any of its employees as to the accuracy of this Pocket Handbook or any information contained in it. No responsibility for loss occasioned to any person acting or refraining from action as a direct or indirect result of any material included in this publication is accepted by the publisher, editors or contributors, and all liability for any such loss is hereby disclaimed to the fullest extent permitted by applicable law. This publication is produced for general information only, and specific advice should always be sought for individual cases. Library and research facilities are (subject to applicable conditions) available at Watts and Partners to assist in developing particular lines of enquiry. Where opinion is expressed by any contributor, it is the opinion of the contributor and does not necessarily coincide with the editorial views of Watts and Partners.

ISBN 1-870776-22-4

© Watts and Partners 2005. All rights reserved. Without limiting the rights under copyright reserved above, no part of this publication may be reproduced, stored in or introduced into a retrieval system, or transmitted, in any form or by any means (electronic, mechanical, photocopying, recording or otherwise) now known or hereafter devised, and whether or not transiently or incidentally to some other use of this publication, without the prior written permission of the copyright holder.

Warning: The doing of an unauthorised act in relation to a copyright work may result in both a civil claim for damages and criminal prosecution.

Published by Watts and Partners
Typeset and printed by Cantate Print, Battersea, London
Cover design work by Art and Industry, Clerkenwell, London

Contents

Introduction	1
Property acquisition	3
Commercial and industrial property	3
Commercial/industrial surveys	4
Single project insurance	6
Due diligence	7
Due diligence and the building survey	8
Vendor's surveys	10
Identifying the age of buildings	11
Development monitoring	12
Housing and residential property	17
Residential surveys	18
Development and procurement	21
Contracts and procurement	21
Public Procurement, Public Private Partnerships and the Private Finance Initiative	22
Procurement methods	29
Procurement and standard form contracts	29
The contract administrator's role	31
Contract management	33
Employer's agent	34
The quantity surveyor's role	35
Partnering	35
Development	37
Construction management	38
Project management	39
Quality management and professional construction services	40
Legislation	43
Building legislation and control	43
Acts of parliament and regulations	44
The Fire Precautions Act 1971	48
Fire Precautions (Workplace) Regulations 1997 and (Amendment) Regulations 1999	49
The Disability Discrimination Act 1995	51
Building and construction regulations	57
The Building (Amendment) Regulations 2004	58
Airtightness	58

Contents

Health and safety at work for surveyors	59
Construction (Health, Safety and Welfare) Regulations 1996	61
The Construction (Design and Management) Regulations 1994	63
The Workplace (Health, Safety and Welfare) Regulations 1992 - Regulation 14	65
Workplace Regulations - provision of sanitary facilities	66
Town and country planning in England	**69**
Development applications and fees	70
Appeals and called-in applications	70
Development plans and monitoring	70
General Development Order and Use Classes Order	71
Planning policy guidance notes and circulars	72
Listed buildings and conservation areas	73
Environmental impact assessments	75
Legal and lease	**77**
Dilapidations	78
Section 18(1) of the Landlord and Tenant Act 1927	83
Latent Damage Act 1986	85
Discovery of building defects - statutory time limits	85
Expert witness	86
Dispute resolution	88
The Provisions of Part II of the Housing Grants, Construction and Regeneration Act 1996	89
Adjudication under the Scheme for Construction Contracts - how to get started	92
Neighbourly matters	**95**
Rights to light	96
Daylight and sunlight amenity	101
Party wall procedure	102
Access agreements	105
Construction noise and vibration	106
Design	**109**
Architectural and design criteria	**109**
Basic design data	110
Building types	111
Specifications	115
Different specifications for different contracts	115
Specification writing	115
Site archaeology	116
Coordinating project information	118

Contents

Structural and civil engineering design	**121**
The design process	122
Soils and foundation design	122
Soil survey	123
Design loadings for buildings	126
Floors in industrial and retail buildings	130
Building services design	**135**
Air conditioning systems	136
Plant and equipment	138
Lift terminology	139
Lighting design	140
Data installations	141
Site analysis	**143**
Measured surveys	144
Computer-aided design	**147**
Computer-aided design	148
Tables and statistics	**151**
Conversion formulae	152
Materials and defects	**153**
Materials	**153**
Deleterious materials	154
Asbestos	156
High Alumina Cement concrete	161
Chlorides	163
Corrosion of metals	165
Building defects	**167**
Common defects in commercial and residential properties	168
Problem areas with 1960s and 1970s buildings	170
Defects in concrete	184
Fungi and timber infestation in the UK	185
Troublesome plant growth: Japanese Knotweed	189
Rising damp	191
Rising groundwater	193
Subsidence	194
Testing	**195**
Chemical and physical testing requirements	196
Non-destructive testing	197
Cladding	**199**
Composite panels	200
Curtain walling systems	201

Contents

Mechanisms of water entry	205
Glazing - windows and doors satisfying the Building Regulations	206
Spontaneous glass fracturing	208

Conservation | 211
Contaminated land	212
Radon	215
Energy conservation	217
Environment and specification	218
BREEAM	220

Maintenance management | 223
Primary objectives of maintenance management	224
A systematic approach to maintenance management	224
Condition surveys	227
'Best value' in local authorities	229
Sources of information in maintenance management	230

Cost management | 233
Urban regeneration	234
VAT in the construction industry - zero rating	235
Capital allowances for taxation purposes	239
Land remediation allowances	242
Tender prices and building cost indices	243
Rating revaluation 2005	244
Cost management	245
Life cycle costings	246
Reinstatement valuations for insurance purposes	252

Technical resources | 253
Useful information sources	254
Useful website addresses	255
Contributors to the Watts Pocket Handbook	260
Watts and Partners' publications	264

Index | **266**

Watts and Partners offices | **278**

Introduction

Once again, it is my great pleasure to introduce the Watts Pocket Handbook 2005 - your invaluable guide to property and construction.

In no other single reference can you find so much pertinent data - from neighbourly matters to cost management; building defects to development and procurement; and maintenance management to building legislation and control - the Watts Pocket Handbook has it all.

For over 20 successful years, Watts and Partners has compiled and published this essential reference document, with the invaluable assistance of our contributors. As a result, each comprehensive edition is an indispensable guide, with updates to current best practice and legislation, and new and varied topics each year.

This year, Watts Pocket Handbook owners can register for a range of new benefits. Please visit our Handbook website www.WattsHandbook.com to find out more about the monthly technical updates, and other expert advisory services.

We are confident you will find yourself referring to the Watts Pocket Handbook time and time again - whether you are a student studying for a degree, a project manager on a new office development, or a chief executive of a major property investment company.

By the same token, we would welcome any comments and suggestions you may have for adding to or improving our publication.

Peter Primett
Chairman
Watts and Partners

Commercial and industrial property

Property acquisition

▸ Commercial/industrial surveys   4

▸ Single project insurance   6

Property acquisition

Commercial/industrial surveys

Check lists for establishing brief and extent and method of survey:

Identify the reason for and scope of survey
- acquisition or disposal (vendor's survey)
- general or specific
- adaptation, extension, change of use
- maintenance programme, expenditure plan
- value as an investment
- insurance assessment
- investment or occupation.

Ascertain the scope and degree of detail required
- delivery, access
- floor loadings, gantry loadings
- security
- fire protection
- means of escape
- environmental control and other matters
- special processes, dangerous or toxic materials
- degree of opening up of concealed areas.

Identify energy conservation requirements
- use of specialists for inspecting, testing, etc.

Agree extent of tests/analysis of materials
- asbestos (type, content, dust counts)
- High Alumina Cement
- calcium chloride
- carbonation
- covermeter or other non-destructive tests
- methane gas
- nature of core materials in composite panels
- quality/constituents
- extent and method of sampling
- making good
- extent of services tests including drainage
- electromagnetic radiation.

Ascertain access for inspection
- ladder, cradle, hydraulic hoist, abseil, specialist help required
- public liability insurance and insurance of hire equipment
- special access requirements for confined spaces.

Property acquisition

Ascertain tenure and request relevant documents
- ❖ freehold/leasehold/feuhold
- ❖ copies of leases, deeds and any plans attached
- ❖ advice on effects of repairing covenants, schedules of condition, rebuilding, reinstatement clauses.

Contaminated land
- ❖ possible contaminative uses (but check that professional indemnity (PI) insurance covers this risk).

Ascertain information available
- ❖ history
- ❖ base building specification
- ❖ original architects, engineers, developers, contractors
- ❖ floor plans, other drawings, details
- ❖ maintenance records
- ❖ maintenance personnel
- ❖ collateral warranties
- ❖ building contracts/status of works
- ❖ health and safety file
- ❖ fire certificate.

Consider impact of regulations
- ❖ Disability Discrimination Act
- ❖ Workplace Regulations:
    - fire
    - glazing
    - sanitary provisions
    - lighting
    - protection from falling
- ❖ Construction (Design and Management) (CDM), health and safety
- ❖ Fire Precautions Act.

Establish if costings are required
- ❖ approximate for budget purposes
- ❖ state limitations on costing information.

Establish and confirm restrictions
- ❖ access arrangements
- ❖ restricted, high security areas
- ❖ means of identification, security arrangements
- ❖ potentially hazardous areas

Property acquisition

- ❖ normal working hours/weekend working
- ❖ need to undertake risk assessment.

When surveying property – especially when empty – practitioners should conform to the procedures outlined in 'Practical procedures' under the 'Health and safety at work' section of this handbook.

Single project insurance

When developing a building, institutions will require that third parties are given protection against construction defects and consequential loss. Traditionally this has been achieved by procuring warranties from contractors and the professional team in favour of occupiers, purchasers or other funders. More recently, single project insurance has been introduced. This is first party material damage insurance to fund the early repair of damage caused by latent defects, without proof of fault.

A single policy is purchased, with the development team named within the policy schedule (including contractors and the professional team). If a relevant defect is notified to the insurer within the time limit of the policy (usually 10 years from practical completion), then the insurance company should pay out against the policy before investigating the ultimate source of the liability (the fault). The insurance company then makes a claim (subrogation) against the party at fault.

The policy usually limits claims to the 'structure' of the property, so the definition of the structure should be checked carefully, particularly if the single project insurance policy is replacing warranties, rather than supplementing them. It is also not unusual for services installations to be omitted, unless an additional premium is paid.

The cost of the policy is generally between 1% and 2% of the construction cost, so it is relatively expensive. Additional premiums are sometimes paid to cover the effects of inflation during the term. The insurer will have to make an independent assessment of the development team, the design and the quality of construction. This will usually involve the services of a monitoring surveyor, or engineer.

Single project insurance is also known as: BUILD (building users' insurance against latent defects); latent defects insurance; decennial insurance; wrap-up insurance; or omnibus insurance.

Due diligence

- Due diligence and the building survey — 8
- Vendor's surveys — 10
- Identifying the age of buildings — 11
- Development monitoring — 12

Property acquisition

Due diligence and the building survey

The term 'due diligence' fell into common use following the US Securities Act of 1933. So long as broker dealers conducted a due diligence investigation into a company whose equity they were selling, and disclosed what they found to the investor, they would not be held liable for non-disclosure of information that failed to be uncovered in the process of that investigation.

The building survey is just one part of the process of property acquisition; its significance should not be underestimated. While it may be true to say that the decision to purchase or occupy is often governed more by commercial pressures than by faults in the building, the due diligence process is designed to alert the purchaser to issues that will affect the building as an investment and/or as an asset – to manage the risks that are inherent in property acquisition. The 'conventional' survey may therefore be expanded to encompass a wide range of issues that might, in the normal course of events, be disregarded.

There is no hard and fast rule as to what should be included in the process and what should not; the particular nature of a deal or building will dictate particular investigations. However, the following issues may be relevant:

Occupational considerations

- Constructional issues relating to fit out and occupation
- Suitability of space for particular use
- Ease of sub-division/sub letting
- Contributions from landlord
- Efficiency of space usage
- Quality of base build information and accommodation
- Oversized/undersized
- Access
- Parking restrictions
- Security
- Service charge levels - past history
- Workplace Regulations (fire, lighting, glazing, sanitary provision, protection against falling)
- Disability Discrimination Act
- Occupancy cost review.

Repairs and defects

- Patent defects
- Potential latent defects
- Compliance with statute
- Schedules of Condition (and effects thereof)
- Onerous maintenance issues
- Deleterious materials
- Quality of finishes/construction
- Unusual construction techniques.

Property acquisition

Environmental considerations
- Risk of contamination
- Previous site use
- Remediation
- Power lines
- Japanese Knotweed
- Flooding
- Radon
- Coast erosion
- Mining
- Security.

Legal issues
- Boundaries
- Services crossing boundaries
- Party wall issues
- Rights to light
- Restrictive covenants
- Planning
- Means of escape over adjoining land
- Adjoining uses
- Warranty package
- Building contract issues
- Defects liability
- Title
- Repairing obligations
- Lease break provisions
- Reinstatement obligations
- Dilapidations liability
- Fire certification.

Due diligence team members
- Building surveyor
- Services engineer
- Environmental engineer
- Agent
- Managing agent
- Solicitor
- Planner
- Space planner
- Structural engineer
- Cladding specialist

Property acquisition

- ❖ Project manager
- ❖ IT consultant
- ❖ Cost consultant.

Vendor's surveys

The traditional approach to selling investment grade property is to release details to the market, consider offers and reach heads of terms which are often subject to survey and legal enquiries. This due diligence process may give the prospective purchaser cause to renegotiate terms with a risk to the vendor in terms of cost and time. If matters proceed smoothly, exchange of contracts and completion can then take place.

The process is time consuming and fraught with risk. To streamline the procedure and manage risks, vendors are increasingly procuring full survey reports prior to sale – the government's 'Sellers Pack' for the residential market is a similar concept. Vendor's surveys can be defined as building and other surveys commissioned by a vendor but primarily for the benefit of a purchaser.

Normally, the vendor's sale pack will include the normal building survey, a phase 1 environmental report or land quality statement, test reports on deleterious materials (where relevant) and probably a report on the building's services installations. For housing transactions, the proposed seller's pack will include:

- ❖ terms of sale
- ❖ evidence of title
- ❖ standard searches
- ❖ planning and building control certificates
- ❖ seller's information form
- ❖ warranties
- ❖ Home Condition Report.

Whereas a conventional report may often make recommendations for further investigation, a vendor's survey must not invite further questions. Thus, it is very important to either make a judgement and express an opinion based upon the evidence, or commission additional tests and inspections where it is relevant to do so. Similarly, questions that would normally be referred to the legal team should be addressed in advance of the production of the final report. In other words, every effort must be made to 'close' particular issues or observations.

Traditionally, building and other surveys include a number of limitations. The third party clause (where the report can only be relied upon by the client) clearly needs variation with vendor's surveys. There is always some scope for discussion, but the usual basis is that the client (or vendor) can rely upon vendor's survey reports as well as the first purchaser. A duty is often extended to the first purchaser's bankers. Furthermore, there should be no change of use, as this may have impacts on the building that the surveyor cannot foresee. The purchaser must also accept that the building's condition may have changed since the date of our report. Assignment of the report to the first purchaser is traditionally done by exchange of letters but can also be executed as a deed.

An advantage of a vendor's survey is making information on the condition of the property available to the vendor, which is essential for good asset management. The surveys also take technical due diligence off the critical path, because they are prepared before the property is marketed. Heads of terms can be entered into without being subject to survey. This approach may also give less scope to prospective purchasers who are not sincere in proceeding on the basis of their original offer.

Property acquisition

Identifying the age of buildings

This guide cannot pretend to be a comprehensive dissertation on the architectural and construction history of Britain. It is a concise yet wide-ranging guide to clues and explains what to look for when seeking to assess the age of a building. Always remember that, almost without exception, any building, unless of very recent construction, will, at some time in its life, have undergone some form of change, modernisation or conversion that may well hide the age of the original construction.

If you have the time and opportunity always seek to consult such archival sources as are immediately available. Even if the building is not a historic building, some records of the original date of construction will exist somewhere in the files of the local administration.

If the building is Listed as being of Special Architectural or Historic Interest or a Scheduled Ancient Monument the listing description or entry in the Scheduled Monuments Register will provide some indication of the believed age of the building or monument. A word of caution here is that these believed ages are based in most cases on external inspections only and even the most experienced inspectors of historic buildings and ancient monuments have been known to have been deceived. This usually takes the form of putting a more recent age on the structure based on external elevations that are the result of, say, a late Victorian refronting of a Georgian Building, or an early 18th century brick refronting of a Tudor or earlier timber framed building. This last is often referred to as a Queen Anne front on a Mary Anne back, therefore, always ensure that you look at the back of the property as well as the front.

Where to go to find records

Every county in Britain has a County Archivist and contact can be established through the local authority of the area in which the building is situated. Alternatively contact can be made via the relevant heritage authority - English Heritage, Scottish Heritage, Welsh Heritage [Cadw] or the Northern Ireland Heritage Department of The Department of Environment of Northern Ireland. Alernatively the Local Authority Conservation Officer will be able to put you in touch and indeed the relevant conservation or heritage authority or office could be a useful source in their own right.

Archives, be they county or district or more local, down to even Parish level, will include one or more sets of the following which may assist:-

- ❖ Maps especially tithe maps showing down to considerable detail ownerships and the building in outline on each site or plot.

- ❖ Sale documents, especially auction details that themselves give indications of believed age[s] of the building.

- ❖ Newspaper and other articles indicating the believed age[s] of the building.

- ❖ Deeds Registers, which can be particularly useful in establishing exact dates for the original building lease from the lord of the manor or landed estate owner, granting the right to the construction of the building being considered.

- ❖ Local authority building bylaw and drainage permissions or approvals to the construction of the building and subsequent alterations. This last one usually only reaches back to the mid 19th century but the equivalent landed estate or manorial records can reach back much further. If the building is still held in freehold or equivalent by the landed estate or manorial estate then such records may still be held by them. However, some such estate records have been transferred to the county or district archives.

Property acquisition

Published and unpublished archival research

The Victoria County Histories, which should be available in the county or district central reference library may contain a reference to your building.

London has the Survey of London volumes now produced by the Survey of London branch of English Heritage. Those started in the late 19th century only cover a part of the historic areas of London. Again available in principal reference libraries.

Other public and private publications may exist for your particular building. The Dr Nikolaus Pevsner 'Buildings of England/Britain' series may assist.

Local history librarians are a mine of information in such a search. They and the county or district archivists also may be able to assist in pointing you towards unpublished works either held by them or produced by local historians.

Research in specialist public archives can also be extremely positive in producing plans and documentation. For any building that is or has been in government or Crown ownership, the Public Record Office at Kew can be invaluable, with large parts of their catalogues available online.

See www.nationalarchives.gov.uk and enquiry@nationalarchives.gov.uk

For a building by an important architect the V&A/RIBA Drawings Collection now [as from 18 November 2004] at the Victoria and Albert Museum can produce drawings back to the seventeenth century.

For London, the Metropolitan Archives [Tel 020 7332 3820] hold papers dating back to the sixteenth century.

For any building that has at any point in its life been in the direct ownership of, or occupation by the Monarchy the Royal Archives at Windsor Castle can produce vital details unobtainable elsewhere.

The building

Once you have understood the types and forms and materials of construction used in particular periods, as you inspect or survey a building the less altered areas of the building can be very revealing. This is particularly true of roof spaces, basements, rear elevations, back or rear additions or anywhere else that has escaped the "improver of antiquity" ie the previous owner[s] who modernised the property. As you crawl around the building look out for such unaltered places.

Above ground archaeology is the increasingly common term for such on-the-property investigations.

Development monitoring

The need for development monitoring

Development monitors are appointed to oversee a third party's interest in a development. Development monitoring involves identifying, advising on and monitoring construction related risks, which are not under the client's direct control. There is a need for the monitor to actively protect the client's interest during the lifetime of the development. This is particularly important where the client takes a more risk-averse approach to the project than the developer.

A key element of the role is that advice is provided to the client from a party that is independent and not associated with the development or the development team.

The need for development monitoring is diverse and each instruction will reflect the requirements of the client and their relationship to a particular development. Development monitoring is typically undertaken on behalf of investment funds, banks and future occupiers, as summarised in the following table.

Property acquisition

Type of organisation	Examples of organisation	Interest in development
Fund	Property investment fund, pension fund, private equity fund, joint venture partner	Will purchase the scheme as an investment on completion, or will acquire the land and fund the development during construction as an investment
Funder/lender	Clearing banks, investment banks	Will earn interest on a loan for the land purchase and/or the building during construction, as well as a pre-agreed arrangement fee and possibly an exit fee
Prospective occupier	Various tenants	Will acquire a leasehold interest in the completed development, which may include full repairing and insuring obligations

The monitoring service

The development monitoring service will depend on the nature of the client's interest in the development and the risks associated with this interest. Therefore there is no standard service. The client and the monitor must work closely together to ensure the monitor's brief fully meets the specific project requirements.

Where a funder is providing debt finance, the principal focus is likely to relate to the value of the works undertaken and the completion of the works within the agreed budget. Progress of the works during construction will also be an important aspect to monitor. However, where the loan is secured against the property, it is prudent for the funder to ensure that the quality of the works is also considered.

In situations where the client has, or has agreed to, purchase the development, a quality focus is more prevalent. The client will want to ensure that the completed development is of a quality suitable as an investment asset. Where the fund is financing the works, it is common for development and purchase agreements between funds and developers to include maximum cost limits for the fund, or developer profit erosion provisions, which protect the fund from cost over runs.

Progress of the works may be critical if there is an agreement for lease with a future tenant, or tenants. In this instance, close monitoring of the likelihood of achieving the access date(s) for the tenant's fit out works will be required. This can be undertaken on behalf of the fund, funder or future occupier.

In general terms the monitor will comment on the cost, programme and quality of the proposed development, from the initial concept to full design, and throughout the construction period, and into the defects liability period.

The key skills and competencies that should be sought in a development monitor are:

Property acquisition

- Diverse technical knowledge of construction
- Thorough understanding of the client's interest in the development, including the development/purchase/finance agreement and the conditions precedent
- Project management and procurement expertise
- Cost management knowledge
- Risk awareness
- Independence from the development team
- Reporting and appraisal skills
- Proactive and a good communicator.

Development monitoring is usually carried out in four stages. Each stage is listed below and is discussed in the following sections.

- Appraisal and risk assessment
- Construction and finance monitoring during the works
- Advice at practical completion
- Advice at end of the defects liability period.

Appraisal and risk assessment

The development monitor will request and then review a wide range of information on the proposed development, including the programme, construction cost, design quality, future maintenance requirements, procurement route, selection of consultants and contractors, building contract, provision of warranties, insurances, statutory requirements, neighbourly matters, and land contamination issues. This review is summarised in an appraisal report, which will highlight any potential risks associated with the project.

The focus of the appraisal will reflect the needs of the client and their interest in the particular development. This should be defined in the legal documentation, such as the development/purchase/finance agreement or the agreement for lease. Where possible it is advantageous for the development monitor to provide technical advice to the client on the construction related matters at the pre-agreement stage. Such advice would relate to any potential significant risks associated with undertaking the development that may have an adverse effect on value.

Construction and finance monitoring

Following the appraisal stage and agreement to proceed with the development, monitoring is undertaken during the construction works. This involves periodic site inspections and attendance at site progress meetings, typically on a monthly basis. A formal report will be prepared that comments on a variety of construction issues, including:

- Construction costs, cash flow and expenditure against budget
- Progress of the works against programme
- Quality of workmanship on site
- Development of the design
- Status of any statutory approvals
- Status of appointment documents and warranties
- Status of construction warranties.

Property acquisition

Practical completion and the end of the defects liability period

The role of the development monitor at practical completion and at the end of the defects liability period is usually set out in the development/purchase/finance agreement or in the agreement for lease. A robust role may be specified involving inspection of the completed development and advising the client on whether they should accept the development as complete, or that the end of the defects liability period has been achieved.

Practical completion is particularly critical where a prospective tenant is in place, as practical completion is likely to trigger rent commencement. Furthermore, practical completion is likely to result in payment for the building to the developer (projects which are not interim funded), or payment of the developer's profit.

The monitor can help to ensure that all the necessary handover documentation is in place, such as the health and safety file, as built drawings, the operation and maintenance manuals and warranties.

Housing and residential property

 Residential surveys   18

Property acquisition

Property acquisition

Residential surveys

Suggested pre-survey checklist for surveyors:

Confirm instructions
- nature of the instructions
- date and time of the survey
- access arrangements, particularly occupied premises
- statement of surveyor's intentions
- limitations.

Bear in mind
- Unfair Contract Terms Act 1977
- liability in negligence and contract
- "requirements of reasonableness".

Equipment to take – consider the following
- powerful, robust torch
- claw hammer and bolster
- ladder (minimum 3 metres long)
- pocket probe
- binoculars or telescope (not more than x8)
- hand mirror (minimum 100mm x 100mm)
- moisture meter
- screwdrivers – assorted
- measuring rods or tapes, notebook and writing equipment
- plumbline
- spirit level
- first aid kit
- protective clothing, hard hat and suitable footwear.

Site notes to be recorded
- notes of defects
- record of weather, persons present, etc
- answers to queries of vendor/neighbours, etc.

Suggested outline of report
Introduction
- brief
- limitations
- general description of property and situation
- present accommodation.

Property acquisition

Structural condition and state of repair

External
- roofs
- other defects at roof level
- eaves
- flashings
- external walls
- likelihood of cavity wall tie failure
- airbricks (provision and adequacy)
- damp proof course
- foundations and settlement/subsidence/heave
- rainwater goods
- soil/waste stacks, gullies
- other external comments
- external decorations.

Internal
- roof spaces
- partitions
- plasterwork
- windows
- doors
- joinery
- ceilings
- other internal defects
- internal decorations.

Services
- plumbing/wiring
- heating
- electricity/gas
- sanitary fittings
- drainage.

Outside
- boundaries, pavings, fences, gates and outbuildings
- noise, contamination and other environmental factors
- adjoining properties
- garden/trees – risk of ground movements/mining subsidence
- garage/car parking
- probability of flooding.

General
- compliance with statutory regulations
- planning situation

Property acquisition

- any responsibilities under a lease
- limitations of report
- any appropriate approximate costings
- presence/condition of toxic materials, for example, asbestos
- general condition and conclusion
- risk and likelihood of fungal decay, insect infestation or conditions that could give rise to these attacks
- risk of contaminated land
- risk of Radon emissions, power lines, etc
- any other relevant information.

Golden rules

When drafting the report, endeavour to avoid technical jargon and take care to communicate precisely.

Avoid assumptions and ill-thought out statements.

When describing elements answer the following questions:-

1. What is it?
2. What is wrong with it?
3. What will need to be done to put it right?
4. What are the consequences of not putting it right?

Health and safety

When surveying a property – especially one that is empty – practitioners ought to conform to the procedures outlined in 'Practical Procedures' under the 'Health and safety at work' section of this handbook.

See also 'Surveying Safely' published by The Royal Institution of Chartered Surveyors.

Contracts and procurement

▶ Public Procurement, Public Private Partnerships and the Private Finance Initiative — 22

▶ Procurement methods — 29

▶ Procurement and standard form contracts — 29

▶ The contract administrator's role — 31

Development and procurement

Public Procurement, Public Private Partnerships and the Private Finance Initiative

Public Private Partnerships (PPP) are, broadly speaking, contractual arrangements where the public and private sectors work together to deliver a project or service. This can include many types of arrangement such as transfer of public undertakings to the private sector and private sector investment in public services.

The Private Finance Initiative (PFI) is a type of PPP where the private sector provides an asset-based service under contract to the public sector. The key features of PFI are that there is an underlying asset, often a building or a series of buildings, and that private finance is used to provide the facility and service. In return, the private sector receives a regular payment from the public sector over the life of the PFI concession, typically around 30 years. PFI is one of HM Treasury's preferred forms of procurement.

This chapter describes some of the key features and mechanisms of PFI and public procurement. Because of the complexity of this field, and differences in interpretation and practice between contracting authorities, it should be regarded as a guide only and specific advice sought in appropriate cases.

In May 1997 PFI, as described above, was officially re-named as PPP. However the term PFI is still in common use, and is used in this chapter. The expression PPP is sometimes used to describe a PFI transaction financed jointly by the public and private sectors.

Features of PFI

- ❖ The private sector contracts not only for the provision of the building or other asset, but for maintenance and services over the life of the concession. This encourages the integration of construction and services provision and investment decisions based on whole life cost, rather than capital cost only.

- ❖ The public sector requirements are expressed as outputs (requirements to be met) rather than inputs (standards to be achieved). This allows the private sector to meet requirements by the most cost effective route and encourages innovation.

- ❖ The private sector is incentivised to perform the services by a penalty structure which reduces or suspends the regular payment (called the Unitary Payment) in the event of service failure (or 'unavailability').

- ❖ The public sector remains responsible for the core public service (teaching, healthcare, administration etc.).

- ❖ The PFI is funded entirely by the private sector in return for the Unitary Payment. In some cases, the public sector may offer long stop guarantees to improve Value for Money (see below).

- ❖ The design, construction, operating and maintenance risks associated with the PFI are largely transferred to the private sector. For example, if the project costs more or takes longer than agreed at the outset, the private sector is responsible for the overrun. The assessment, allocation and management of risks is one of the key disciplines in PFI.

- ❖ The transfer of risk enables public sector authorities to benefit from accounting treatment of the Unitary Payment as revenue expenditure, which does not count against their capital allocations.

Development and procurement

- ❖ The PFI proposal must demonstrate Value for Money (VfM), meaning broadly that PFI has to offer a more cost effective solution to requirements than conventional public procurement. This comparison is made via the construction of a Public Sector Comparator (PSC) financial model for the public sector alternative. In some cases a 'shadow bid' is compiled by the public sector as a check on the PFI.

Key advantages and disadvantages

Advantages
- ❖ Public service improvements are brought forward by using private capital to offset the shortage of public capital.
- ❖ Development and operational risk is transferred to the private sector, freeing the public sector to concentrate on the core service.
- ❖ The private sector is incentivised to provide good quality buildings and services and VfM.

Disadvantages
- ❖ The cost of private finance is greater than that of public capital, meaning that the PFI must achieve efficiency improvements to offset this.
- ❖ Major changes in public sector requirements during the concession period may be difficult or expensive to accommodate within PFI.

Projects and clients

Up to July 2003, PFI projects to the value of £35 billion had been signed, and the annual value of transactions is increasing. The largest single category by value is transport, due in part to the inclusion of the London Underground modernisation programme. Other major sectors include health, education and defence.

Types of projects using PFI include headquarters offices, office roll-out programmes, environmental and waste recycling, transport (road, rail and light railway), education, healthcare, prisons, courts, defence, housing and regional projects. Public sector clients include government departments, local authorities, NHS Trusts and government agencies such as the Inland Revenue.

Because of the long-term nature of the transaction and the cost of the tendering process, PFI is rarely used for projects with a value of less than £10 million, and normally they are much larger. A feature of more recent PFI transactions is the grouping of smaller projects into a single PFI to achieve the appropriate minimum size.

The PFI procurement model has also been adopted by some large private or semi-private clients for their building and facilities provision. This is known as 'corporate', or 'private' PFI.

Project inception

The first formal step in the PFI process is the production of an Outline Business Case by or on behalf of the public sector client. This sets out the project brief and business case for the proposal, and forms the basis for applications for PFI credits from central government where applicable. It is followed up with the output specification which sets out the requirements for the project in performance terms, and in due course this forms a key part of the tender documentation.

A key feature of the output specification is that it is expressed in terms of required outputs, rather than prescribed inputs. Within any constraints set by the public sector, the private sector is free to meet these requirements

Development and procurement

by means of its own choosing. Sometimes these depart significantly from those envisaged by the client at the outset.

Transaction structure

The parties to a PFI transaction typically comprise the following:-

- ❖ Public sector client: The government department, local authority or other public sector authority contracting for the project.
- ❖ Private sector partner: Usually a consortium comprising a contractor or developer, a facilities manager or operator, and a bank or other financial institution.
- ❖ Funders: In addition to the consortium partners, external organisations will normally be involved in the provision of funding. These can include banks or other financial institutions, bond issuers, rating agencies and insurers.

A number of companies with an interest in the PFI market have formed specialist subsidiaries or consortia to participate in contracting, funding or operation, particularly in the field of healthcare.

The above parties to the PFI will normally be advised separately and a typical consultant team would include a financial adviser, legal adviser and one or more technical advisers.

The organisational structure of a typical PFI transaction is shown in Table 1 on page 27.

Procurement

Public procurement is subject to EU rules which are intended to promote fair competition across the EU and a transparent procurement process. Some differences in practice will nonetheless be encountered between contracting authorities and member states.

For public works and services above certain cost thresholds, a notice must be placed by the public sector contracting authority in the Supplement to the Official Journal of the European Communities (OJ). Most PFI transactions will exceed the cost thresholds. The cost thresholds are reviewed every two years and those applying from 1 January 2004 are shown in Table 2 on page 27.

The Supplement to the OJ is published daily in electronic form and may be viewed on the internet (see useful website addresses page 239). Tender information services are available on subscription, and a selection of OJ notices are published from time to time in the construction press.

Tendering procedures

There are three types of tender procedure as follows:-

- ❖ **Restricted**

    Expressions of interest are invited and the most suitably qualified are invited to submit a tender.

- ❖ **Open**

    A notice is published and anyone can submit a tender.

- ❖ **Negotiated**

    The contract is negotiated with at least three tenderers, or with the maximum number available. Use of this procedure is subject to justification.

The choice of tendering procedure is made by the contracting authority having regard to the nature of the works or services, and the circumstances under which they are to be delivered. For example, the nature of the service may mean that a specification cannot be drawn up with sufficient precision to allow the use of the Open or Restricted procedure.

Development and procurement

An accelerated procedure can be used where justified in conjunction with the Open or Restricted procedure.

Types of Notice

Prior Information Notice (PIN): optional notice of intent to invite tenders or expressions of interest later. When placed in accordance with the mandatory timescales, it allows the period for submission of tenders or expressions of interest to be reduced.

Contract Notice: main notice providing full information and informing parties how to bid or pre-qualify for the project.

The mandatory timescales for the tender process are shown in Table 3 on page 27.

Content of Notices

Notices are in a form prescribed by the EU and group information into the following sections:-

    I. Contracting authority
    II. Object of the contract
    III. Legal, economic, financial and technical information
    IV. Procedure
    V. [Spare number]
    VI. Other information.

Within these sections, certain information is mandatory. However notices contain more than the minimum information, to encourage a good quality and level of response from bidders.

Certain conventions are used in notices, including CPV (Common Procurement Vocabulary) and NUTS (Territorial Designation). For an explanation of these and other conventions, refer to the SIMAP website (see useful website addresses on page 259).

The official forms for placing of notices in the OJ can be downloaded from the SIMAP website.

Project processes

The post tender processes can vary widely depending on the project nature and external factors. Typically an Invitation to Negotiate (ITN) will be issued to a shortlist of bidders, and this may be followed by a further bidding round (Best and Final Offer or BAFO). In order to reduce bid periods and associated costs, current practice is to encourage the provision of full information by both parties at the outset, and the use of standard contract terms as far as possible.

At the conclusion of bidding and post bid negotiations, a Preferred Bidder is selected, together with a reserve bidder. The parties then negotiate the remaining details of the project and confirm funding, legal and third party issues. During this period, development of the design and construction proposals continues, and planning permission may be sought. Once the terms are fully agreed and funding secured, contract award (or 'Financial Close') takes place, and the project can proceed.

Funding

PFI transactions are mainly debt financed, and senior debt is normally provided independently of the consortium partners. The main sources of PFI funds are:-

- ❖ Equity – normally provided only in limited amounts by the private sector consortium partners.
- ❖ Subordinated or mezzanine debt – provided by consortium partners or other funding institutions.

Development and procurement

❖ Senior debt – normally the bulk of the funding, provided by bank lending and/or via the capital markets.

The choice between bank and capital market options will depend on the nature and size of the project, and the market conditions at the time of placement. Bank debt is suitable for smaller, shorter life or higher risk projects, and the finance cost is generally higher than the capital markets option.

The capital markets option is particularly suitable for larger, lower risk projects for central government, where the covenant is perceived as strongest. This option requires the issue of an index linked or fixed rate bond which is typically taken up by other financial institutions and pension funds. Although the process of issuing a bond is more expensive than arranging bank debt, this will be recovered over the life of the bond in lower finance costs.

The bond issue is underwritten by a financial institution and may be credit rated by an agency such as Standard and Poor's to improve marketability. As an alternative, or in addition, bonds may be enhanced by private or public sector guarantees, or by the use of credit risk insurance (called monoline insurance). A bond enhanced in this way is known as a wrapped bond.

PFI transactions may be re-financed during the concession period, typically by the re-negotiation of debt finance once the development period is over and the project risk profile is reduced. Some PFI contractors have also begun to sell on their equity stakes, creating a secondary transaction market.

Success factors

Some PFI transactions are more successful than others, and some success or failure criteria are subjective. While it is difficult to draw general conclusions, research has identified the six most significant value for money drivers in PFI as: the level of risk transfer; linking the output specification requirements to payment; the length of the contract needed to recoup investment; appropriate performance measures and incentives; a competitive bidding process; and excellent management skills in the private sector.

PFI - an assessment

Since it was inaugurated in 1992, PFI has enabled significant investment in public facilities and services which might not otherwise have taken place. Studies by the National Audit Office have indicated that under PFI, the proportion of projects being delivered late or over budget has fallen from roughly three quarters to a quarter, and that four-fifths of public bodies involved with PFI believe they are achieving satisfactory or better value for money from PFI contracts.

There has been criticism of some early grouped PFI projects on grounds of poor design quality, which has led to a series of design quality initiatives led by the government-sponsored Commission for Architecture and the Built Environment (CABE). Others have criticised PFI on the grounds of doubtful value for money, reduction in safety standards, and dilution of the public service ethos. As the number of PFI projects grows and the process matures, the focus is likely to move from the initial transaction to the quality of FM services and operational management over the life of the concession.

The PFI model, which originated in the UK, has been taken up in several other countries, both in Europe and beyond; and so long as there is a demand for better public services and a continuing shortage of public capital, it seems set to continue and even grow in popularity.

Development and procurement

Table 1 – PFI Transaction Structure

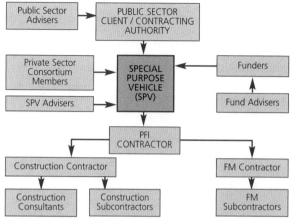

Table 2 – EU Thresholds for Public Procurement

From 1 January 2004 in pounds sterling net of VAT

Public Sector		
	Supplies and Services	Works
Central government bodies	99,695	3,834,411
Other public sector	153,376	3,834,411

Utilities		
	Supplies and Services	Works
Water, electricity, urban transport, airports, ports	306,753	3,834,411
Oil, gas, coal, railway	258,923	3,236,542
Telecommunications	388,385	3,236,542

Note: this is a summary of procurement thresholds only. Reference should be made to the current EU Regulations, which contain exceptions and additional categories.

Table 3 – EU Procurement Process

Key Timescales for Publication in OJ and Return of Tenders
(based on Restricted procedure)

Prior Information Notice (PIN)	Between 52 days and 12 months before Contract Notice.
Contract Notice	Period for expressions of interest 37 days (see Note 1).
Tenders	Period for return of tenders 40 days (26 days if PIN issued as above) (see Note 1).
Contract Award Notice	Within 48 days of award of contract.

Note 1: Accelerated procedure can be used where justified and will allow periods for expressions of interest and return of tenders to be reduced.

Table 4
Glossary of Common PFI Terms

BAFO	Best and Final Offer: final stage in negotiation prior to selection of Preferred Bidder.
Contract Notice	Main notice placed in OJ Supplement to advertise a public contract.
Financial Close	Stage when all contractual and financial terms have been agreed and private sector funding is in place; formal execution of the PFI agreement.
Gateway Reviews	Project stage reviews by the public sector to review and confirm proposals and acceptability of risk profile.
ITN	Invitation to Negotiate: invitation to the private sector to provide detailed proposals leading to negotiations and a decision on the Preferred Bidder.
KPI	Key Performance Indicator: definition of performance requirement(s), failure events and penalties (deductions from the Unitary Payment).
OBC	Outline Business Case: the PFI project brief and business case, produced by or on behalf of the public sector contracting authority.
OJ	Official Journal of the European Communities (also OJEC or OJEU). The OJ Supplement carries notices of public contracts.
Output Specification	Specification of requirements for the PFI project and service, produced by or on behalf of the public sector contracting authority.
PFI	Private Finance Initiative
PIN	Prior Information Notice: a preliminary notice placed in OJ to advertise a public contract.
PPP	Public Private Partnership
Preferred Bidder	Stage in negotiation where a Preferred Bidder (and usually a reserve bidder) are selected to negotiate the detailed terms leading to Financial Close.
PSC	Public Sector Comparator: a benchmark of the cost of the PFI against the publicly financed alternative.
SPV	Special Purpose Vehicle: the company created and owned by the private sector consortium and which contracts for the PFI project.
Unavailability	Service failure against KPI availability criteria, resulting in a financial penalty to the private sector.
Unitary Payment	The regular user payment made by the public sector client to the private sector SPV.
VfM	Value For Money: the objective of best value, or meeting user requirements at the lowest whole life cost.

Development and procurement

Procurement methods

Essentially the function of construction contracts is to assign appropriate levels of risk to those parties best able to deal with them.

Risk is an inherent element of any construction project and therefore risk management is an essential part of contract strategy.

There are currently three main procurement options used within the industry which reflect various ways by which risk is balanced between the parties.

Traditional procurement

A traditional procurement route may be defined as one where the design is largely complete before either the main contractor, sub-contractor or specialist contractors become involved. This may be appropriate for some projects if the client's objectives have been clearly and comprehensively determined. However, it does carry with it the disadvantages of increased economic uncertainty and limited opportunities to refine design and improve cost efficiency. It is also possibly more likely to lead to disputes. Increasingly, when traditional contracts are used, the contractor is selected on a two stage basis to gain some of the advantages that are available from early contractor involvement.

Design and build procurement

One alternative to traditional contracting is design and build procurement, where part or all of the design development and financial outcome responsibility risk is delegated to the constructor. There are many considerations that need to be carefully evaluated in choosing this procurement route, and opinion is divided as to whether it should be used on all types of construction projects. Most commonly this route is used in the procurement of 'spec' buildings, such as industrial units or office buildings when the control of design criteria and building techniques is not a dominant factor for the client.

Management procurement

The key feature of any management contract is that a manager is appointed with the responsibility to manage a project, not just provide advice or consultancy services. Two main derivatives of this form are management contracts and construction management.

Procurement and standard form contracts

Standard form contracts have always been a feature of the construction industry. A summary of the main standard forms relevant to commercial developments is set out below.

Professional appointments

- **Architect** – RIBA SFA/99 (revised April 2002).
- **Engineers** – ACE conditions of engagement 2002. A suite of conditions to accommodate particular engineering services, including also a short form agreement and sub-consultancy agreement.
- **Quantity/Building Surveyor** – **RICS Form with guidance notes 1999** for quantity surveying services and **RICS Form with guidance notes, 2nd edition 2000** for building surveying services.

Development and procurement

- **Planning Supervisor** – The **RIBA, RICS** and **ACE** standard form appointments make provision for planning supervisor services. It is common for the architect, quantity surveyor or engineer to also undertake the role of planning supervisor. Sometimes an independent consultant is appointed to undertake the planning supervisor role.
- **Project Manager** – RICS Project Management Memorandum of Agreement and Conditions of Engagement 3rd Edition September 1999 and RIBA Form of appointment for Project Manager PM99 (revised April 2002).
- **General** – The ICE Professional Services Contract, 2nd Edition 1998. This is part of the New Engineering Contract.

Construction contracts

- **JCT Major Project Form (2003).**
- **JCT 1998 With Contractor's Design** (with amendments 1-5). The JCT design and build form of contract.
- **JCT 1998 Private with Quantities; Private without Quantities; and Private with Approximate Quantities** (the latest forms also incorporate amendments 1-5). The most commonly used traditional building contract forms. There are supplements to provide for sectional completion and contractor's design, which may be incorporated. There are also local authority versions of these forms.
- **JCT 1998 Intermediate Form of Building Contract (with amendments 1-5).** For medium value uncomplicated projects. There is no provision for contractor design responsibility except for specialist sub-contractors (under the prescribed form and tender arrangements) who may be required to give a direct design warranty to the employer.
- **JCT 1998 Minor Building Works Form (with amendments 1-5).** For simple lower value minor works with no design responsibility.
- **JCT 1998 Management Contract** (with amendments 1-5). The employer contracts with the management contractor who in turn contracts with the trade contractors. The management contractor has limited liability.
- **JCT Construction Management Contract.** The employer contracts directly with the trade contractors. The construction manager project manages under a separate contract with the employer.
- **JCT Non-Binding Partnership Charter for Single Projects.**
- **The ICE New Engineering Contract.** This comprises the Engineering and Construction Contract, 2nd Edition 1995 including options for a Priced Contract, a Target Cost Contract, a Cost Reimbursable Contract and a Management Contract. A new 3rd Edition is in consultation and is expected Spring 2005.
- **The NEC Partnering Option X12, 2001.**
- **The ACA standard form contract for Project Partnering, PPC2000** (amended 2003).
- **The ACA standard form of Specialist Contract for Project Partnering, SPC2000.** This is a specialist subcontractor version of the PPC2000 form of contract. It can be used with PPC2000 or on its own.
- **The Be Collaborative Contract (2003).** A partnering contract with a defined 'overriding principle'.
- **Sub-Contracts** – standard forms of sub-contract commonly used are the JCT DSC/C subcontract form 2002 edition which

Development and procurement

is replacing the **Construction Confederation Dom/1** form and **Dom/2** which will be replaced by the new JCT DSC with design form, and the **Engineering and Construction Subcontract, 2nd edition 1995.** A new JCT short form subcontract was published in Autumn 2003 for minor works. The **sub-contract for the Major Project Form** was published in 2004. This can be used as a stand alone subcontract

The JCT Major Projects Form published in June 2003 was a major departure from the traditional style and drafting of JCT 1998. It has been enthusiastically received for use with substantial developments and experienced participants. There is a substantial risk transfer to the contractor who has design and construction responsibilities. A sub contract for this form was published in 2004. The NEC Partnering Option builds on the ECC suite of contract conditions, which were, at the time of their publication, a fresh approach to the drafting of contract conditions. These conditions are currently under consultation for publication of a 3rd Edition expected in Spring 2005.

The ACA PPC2000 adopts a radical approach as the first multi-party standard form contract.

The Be Collaborative Contract was published in 2003 aimed at delivering a collaborative partnering approach. The conditions have a stated 'overriding principle' setting out the approach required by the parties.

The contract administrator's role

It is a feature of most construction contracts that a person is appointed by the employer to administer the terms of the contract on the employer's behalf.

The contract administrator owes a duty of care to the employer. Under the terms of the contract he must undertake a number of administrative functions including the following:-

- ❖ Managing the client/contractor.
- ❖ Coordinating the pre-project, project and post-project phases.
- ❖ Instigating client variations.
- ❖ Agreeing interim payments.
- ❖ Issuing certificates including payment certificates and practical completion.

The contract administrator has an important role in giving advice and information and also monitoring the work. However, he/she must also remain unbiased in matters such as certification of payments and ensuring that the contract terms are adhered to.

The mandatory nature of these duties is reflected in the contract between the contract administrator and the employer and in contract between the employer and the contractor. As such, there is often considerable scope for disagreement between the contracting parties, both in contract and tort, on whether these duties have been satisfactorily performed.

It is with this background, that those fulfilling the contract administrator's role should be clear on what is required of them.

As a contract administrator the following pre- and post-contract services should be considered important.

Pre-contract

- ❖ Agree detailed brief with client. Agree procedures for modifying the brief as commission proceeds and recording variations.
- ❖ Set budget and project time constraints and establish reporting procedures.

Development and procurement

- ❖ Establish clear routes of responsibility for design, drawings and specification and ensure client approval as scheme develops.
- ❖ Prepare tender documents and approve tendering processes with client, including the selection process.

Post-contract

- ❖ Administer the terms of the building contract during operations on site.
- ❖ Inspect the progress and quality of the work on a regular basis.
- ❖ Prepare interim financial reports to client including the effects of any variations.
- ❖ Agree interim valuations with the contractor in accordance with the contract allowing for the submission of interim applications for payment.
- ❖ Convene and chair site progress meetings on a regular basis to principally discuss progress, cost and quality issues. Record all principle matters in minutes distributed to all project team members.
- ❖ Prepare regular progress reports for client, including details of any applications for extensions of time or disputes.
- ❖ Agree practical completion of the works under the contract terms. This may include the preparation of 'snagging' lists detailing the non-completion of minor work items remaining outstanding at practical completion.
- ❖ Administer the contract conditions during the defects liability period ensuring that all items of disrepair are rectified before the issue of the making good defects certificate and release of retention monies.

Contract management

- Employer's agent — 34
- The quantity surveyor's role — 35
- Partnering — 35

Development and procurement

Development and procurement

Employer's agent

Employer's agent duties
Increasingly, clients wish to procure projects in the pursuit of certainty of cost. This has led to the rise in the various forms of design and build procurement which transfer the 'risk' in any development to the contractor, while giving him greater flexibility to deliver the product. As an agent acting on behalf of the employer, it is essential that the following pre- and post-contract services are provided:

Pre-contract service
- Define the responsibilities of the employer, employer's agent and contractor
- Appraise and quantify the risks
- Formulate the employer's brief and identify specific requirements
- Assess the contractor's proposals and ensure compliance with the employer's requirements
- Undertake design audit of the contractor's proposals for compliance with the employer's requirements
- Evaluate the offer, the contract sum analysis and stage payments and assess value for money.

Post-contract service
- Set up quality control procedure and report on works carried out on site
- Provide site visits and chair meetings
- Implement changes to the employer's requirements only on written approval of the client
- Agree stage payments and recommendations for payments
- Prepare monthly project control statements and cash flow forecasts to client
- Advise on practical completion, preparation of snagging schedules and component literature.

General exclusions
- Checking and verifying contractor's design in terms of adequacy and efficiency
- Checking and verifying contractor's design in terms of fitness for purpose.

In undertaking duties as the employer's agent, it is important to recognise the contractor's freedom to design, while respecting the client's brief and auditing the quality of the end product.

Development and procurement

The quantity surveyor's role

Capital project advice

The function/mission of the quantity surveyor may be defined as the optimisation of purchasing from the construction industry.

Several capabilities are required to fulfil the mission:-

- ❖ An ability to predict future costs from limited information and in dynamic market conditions.
- ❖ An ability to manage the procurement process to ensure that predictions of cost, time and quality are delivered.
- ❖ An awareness of risk with a capability to assess and manage that risk.
- ❖ An ability to demonstrate value for money.

Current trends in the industry lean towards cost reductions, with pressure coming from major purchasers; the importance of value for money is undoubtedly here to stay. Fixed out-turn costs are a necessity.

This is against a backdrop of upward cost pressure on suppliers of construction resources. Skills shortages are emerging and price rises are sticking.

The challenge to the construction industry and the quantity surveyor is sizeable. Changes in working practices and culture are inevitable and there needs to be a significant change in emphasis towards:

- ❖ value rather than cost;
- ❖ more professionalism in the early stages of project delivery;
- ❖ effective team work between the construction professions and contractors – adversarial relationships are no longer appropriate and will not survive;
- ❖ an understanding of the clients' business, and with that, the ongoing costs of the finished project;
- ❖ an awareness of financial incentives and opportunities available to the client (VAT relief and capital allowances – for details, see relevant sections in the cost management and taxation chapter); and
- ❖ an awareness on the part of purchasers that effective management of the construction process is a valuable service and does not come cheap.

Partnering

The traditional approach to construction procurement is driven by the terms of a contract. Relationships are imposed rather than developed; cost is focused at the expense of value; problems are packaged into liabilities rather than accepted as joint responsibilities; risk is off-loaded rather than managed. Partnering offers an alternative.

Partnering unites the sponsoring, design and construction teams and it endeavours to drive them forward with a common purpose.

The important characteristics are involvement, ownership and trust. This becomes obvious when one considers the facts:-

- ❖ The parties' objectives are not mutually exclusive.
- ❖ A construction project should not comprise a discrete stage of design followed by construction.

- ❖ Generally, parties to a venture perform better if given a measure of control and a share in the action. This is all about ownership and being a stakeholder.

There are several levels of partnering. The simple indicators include the following:-

- ❖ Early involvement of contractors and subcontractors; viewing a construction team rather than a design team and a contractor.
- ❖ Flexibility – for example, being prepared to accept higher short-term costs in the interests of longer-term value; reducing cost rather than simply transferring it to others.
- ❖ A willingness to look beyond contractual responsibilities and provide more than is called for. 'The gold service' – going the extra mile.
- ❖ Concentration upon relationships rather than contractual positions. Actions driven by a common purpose rather than a book of rules.

More formal arrangements are implied in:

- ❖ project partnering; and
- ❖ strategic partnering.

The tools implied in partnering usually involve the following:-

- ❖ Two stage tendering – getting the contractor on board early.
- ❖ Negotiation – allowing the contractor to contribute to the design process rather than having to react to it; gaining a clear understanding of the project's objectives before committing to contract and ultimately being able to reduce his/her costs and, in that process, the client's costs. Perhaps more importantly, however, if the process of negotiation is conducted correctly the contractor is afforded the opportunity of understanding the client's definition of value.
- ❖ Open book – everything competitively tendered but with the leader in the procurement process, the contractor, fully acquainted with the project's decision-making rationale.
- ❖ A charter – as distinct from the rigidity of a contract, the parties set out what they want to achieve rather than what they are obliged to do.
- ❖ Performance related reward – allowing the contractor to become a stakeholder in the project budget rather than a conduit through which it flows.

Ultimately of course, partnering is all about attitude. The logic of partnering is inescapable; the extent to which it works will depend on the ability of the parties to shake free from traditional attitudes but also on the ability of the management team to deliver a project in unison and without (necessarily) conventional contract protection.

Development

- Construction management — 38
- Project management — 39
- Quality management and professional construction services — 40

Development and procurement

Construction management

Under this form of procurement, the client appoints a construction manager who is paid a fee for managing and overseeing pre- and post-contract project activities. Site overheads may be included or paid direct by the client. Separate direct appointments will generally exist for the design team members. Building can begin as soon as the design is sufficiently advanced to allow the initial stages of construction to proceed and to prepare and agree a cost estimate for the entire scheme.

Construction is divided into works packages – either on single or multi-trade lines – and the client enters into direct contracts with each of the individual package contractors. This differs from management contracting where the contracts are between the management contractor and each package contractor. The individual packages are tendered at appropriate times throughout the construction period under the direction and management of the construction manager.

Whether a given project is suited to construction management (CM) will depend on a number of factors, as identified in the following table.

Size	CM is an involved form of procurement and can seldom be justified for small projects (less than £1 million certainly and generally much larger).
Complexity	CM is suited to complex projects with a substantial proportion of specialist package contractor involvement.
Uncertainty	Where the project is being conducted in an uncertain environment, CM affords clients a higher degree of flexibility to make changes during the process while minimising the consequent time and cost penalties.
Time	CM permits overlap between design and construction because tendering of packages can take place on a staggered basis rather than all at once as per traditional arrangements. This can save time and suit a 'fast track' approach. Greater involvement of the client with the CM and package contractors may also improve efficiency and introduce time savings. It should be noted however that time savings may be at the expense of cost risk as the scope isn't fixed until much later in the process.
Cost	Certainty of final cost is possible if the scope is finalised before the package contracts are placed. Obviously this sacrifices some of the flexibility referred to. This basis of contracting (in common with management contracting) should result in the lowest cost at the end of the day because the best price is chosen for each package contract without a main contractor's risk provision. Clients have more options to alter or adapt the design throughout the construction period without leaving themselves at a negotiating disadvantage with a main contractor. There is also greater potential to 'value engineer' the project under these arrangements.
Design	Often clients wish to retain control of the design process and this is facilitated under CM. Changes to design during construction carry less cost and time risk than under other procurement methods.

Development and procurement

Project management

Project Management can be defined as the management of any complex activity that has a discrete beginning and an end. It therefore has a direct application to all industries and particularly to the construction process. The application of project management techniques in construction has developed rapidly over the past 25 years in response to the growing complexity of projects and the need to coordinate an increasingly fragmented and specialised industry.

The first task of a project manager is to define precisely the scope of the project that he or she is tasked with managing, including gaining an understanding of the fundamentals of the client's business plan for the scheme.

The second task is to gain an understanding of the client's objectives in terms of time, cost and quality. These objectives should be specified by way of target dates, budgets and specification standards respectively. The relative priority that the client gives to each of these measures should also be understood. For instance, a client may have an overriding need to complete by a certain date, even to the extent of incurring extra costs to make this more certain. The project manager needs to reflect these factors in a strategy and plan for the project, usually termed the Project Execution Plan (PEP).

The third task is to assemble a team to deliver the project. The team needs to be briefed on the client's objectives and their precise role in the project, and coordination and reporting routines should be defined. An effective project manager will be skilled in leading and directing a team drawn together specifically for the duration of the project. This requires personal, presentational and, above all, communication techniques.

Stimulation of co-ordination in the project team and control of time, cost and quality requires the establishment of master control documents, such as the PEP project programme, cost and quality plans. These are invaluable tools to enable the project manager to promote co-ordination, monitor progress, evaluate status and manage delivery.

Successful construction project management entails:

- strong leadership;
- an empowered client project sponsor;
- project team focus on client's objectives;
- strong team relationships and motivation;
- a strong business case;
- a clear and developed brief;
- identification of measurable time, cost and quality objectives;
- defining roles and communicating objectives throughout the team;
- establishment of strategic as well as detailed controls;
- recognition that only the remaining time, cost and performance can be managed;
- management of risk;
- robust project management practice - alive to risks and opportunities;
- an appropriate procurement and delivery methodology;
- an appropriate project contractual matrix;
- balanced risk apportionment;
- early establishment of control documentation;
- dynamic programme management and control;
- establishment of gateway reviews at key project stages;

- ❖ open and transparent lines of communication;
- ❖ continual challenge to ensure best quality, long-term value and programme;
- ❖ being alive to risk and value; and
- ❖ active management.

Quality management and professional construction services

ISO 9001:2000 has superseded the previous 'triple pack' of ISO 9001:1994, ISO 9002:1994 and ISO 9003:1994. To maintain accreditation, organisations whose quality management systems were compliant with the 1994 standard had until 15 December 2003 to undergo a transition to ensure compliance with the 2000 standard.

The revised title, 'Quality management systems - requirements' reflects the change from assuring quality to producing a quality management system. The requirements of the 2000 standard are more generic in nature and it is intended to be more relevant to all categories of product and professional services and size of organisation.

ISO 9001:2000 is intended to be simpler and more flexible for organisations to adopt and use. The main difference is that it is process driven and a management system standard which integrates quality management with business management.

The content has been arranged into five main sections instead of the original 20. The five sections are identified as follows:-

Quality management system. An organisation needs to establish what its processes are, how they interact, what resources are required to provide the product and how the processes are measured and improved. Once this has been established then a system for the control of documentation has to be established along with the Quality Manual and controls for looking after records.

Management responsibility. This requires management to set policies and objectives, to review the systems and to communicate to the organisation the effectiveness of the system.

Resources management. An organisation needs to identify the resources it needs to ensure that the customer receives what has been agreed. This means not only people but also the physical resources such as equipment, premises and any support services required.

Production realisation. This deals with the processes required to deliver to the customer the product/services they require . These processes cover activities such as taking the instruction from the client, the design and development of the particular service/product, the purchasing of services and materials, and the delivery of the services and products.

Measurement analysis and improvement. This is the measurement and monitoring of the management system, the products and services and customer satisfaction and the analysis of data for the continuing improvement of the system.

The implementation of BS EN ISO 9001:2000 is on the basis of the 'plan-do-check-act' principle:

Plan

- ❖ Identify customer needs and expectations;
- ❖ Strategic planning; and
- ❖ Set policies and objectives.

Development and procurement

Do
- ❖ Implement and operate the processes.

Check
- ❖ Collect business results;
- ❖ Monitor and measure the processes; and
- ❖ Review and analysis.

Act
- ❖ Continually improve process performance.

Building legislation and control

Legislation

➡ Acts of parliament and regulations   44

➡ The Fire Precautions Act 1971   48

➡ Fire Precautions (Workplace) Regulations 1997 and (Amendment) Regulations 1999   49

➡ The Disability Discrimination Act 1995   51

Legislation

Acts of parliament and regulations

England and Wales

There are hundreds of Acts relating to the design, construction and maintenance of buildings in England and Wales. Listed below are some of the more relevant.

The Access to Neighbouring Land Act 1992
The Agriculture (Miscellaneous) Act 1968
The Ancient Monuments & Archaeological Areas Act 1979
The Arbitration Act 1996
The Betting, Gaming and Lotteries Act 1963
The Betting, Gaming and Lotteries (Amendment) Act 1985
The Building Act 1984
The Building Regulations 2000
The Building (Amendment) Regulations 2001
The Building (Amendment) Regulations 2002
The Building (Amendment) (No 2) Regulations 2002
The Building (Amendment) Regulations 2003
The Building (Amendment) Regulations 2004
The Building (Amendment) (No 2) Regulations 2004
The Building (Approved Inspectors) Regulations 2000
The Building (Approved Inspectors etc) (Amendment) Regulations 2002
The Building (Approved Inspectors etc) (Amendment) Regulations 2003
The Building (Approved Inspectors etc) (Amendment) Regulations 2004
The Building (Prescribed Fees) Regulations 1994
The Capital Allowances Act 1990, 2001
The Celluloid and Cinematograph Film Act 1922
The Chronically Sick and Disabled Persons Act 1970
The Chronically Sick and Disabled Persons (Amendment) Act 1976
The Civil Procedure Rules 1998
The Clean Air Act 1956, 1968 and 1993
The Coal Mining Subsidence Act 1991
The Commonhold and Leasehold Reform Act 2002
The Construction (Design & Management) Regulations 1994, 2000
The Construction (Health, Safety and Welfare) Regulations 1996
The Contaminated Land (England) Regulations 2000
The Contaminated Land (England) (Amendment) Regulations 2001
The Contaminated Land (Wales) Regulations 2001
The Contracts (Rights of Third Parties) Act 1999
The Control of Asbestos at Work Regulations 1987
The Control of Asbestos at Work (Amendment) Regulations 1998
The Control of Asbestos at Work Regulations 2002
The Control of Lead at Work Regulations 1998
The Control of Lead at Work Regulations 2002
The Control of Substances Hazardous to Health Regulations 1999
The Control of Substances Hazardous to Health Regulations 2002
The Control of Substances Hazardous to Health (Amendment) Regulations 2003
The Control of Pesticides Regulations 1986
The Control of Pollution Act 1974
The Control of Pollution (Amendment) Act 1989

Legislation

The Countryside and Rights of Way Act 2000
The Defective Premises Act 1972
The Disability Discrimination Act 1995
The Disability Discrimination Act 1995 (Amendment) Regulations 2003
The Disabled Persons Act 1981
The Education (School Premises) Regulations 1999
The Electricity Act 1947, 1957, 1972, 1989
The Electricity at Work Regulations 1989
The Energy Act 2004
The Environmental Protection Act 1990
The Environment Act 1995
The Factories Act 1961
The Fire Precautions Act 1971
The Fire Precautions (Workplace) Regulations 1997
The Fire Precautions (Workplace) (Amendment) Regulations 1999
The Fire and Rescue Services Act 2004
The Fire Safety and Safety of Places of Sport Act 1987
The Food Safety Act 1990
The Gaming Act 1968
The Gaming (Amendment) Act 1980, 1982, 1986, 1987 and 1990
The Gas Safety (Installation & Use) Regulations 1998
The Health & Safety (Safety Signs & Signals) Regulations 1996
The Health & Safety at Work etc. Act 1974
The Health & Safety (Miscellaneous Amendments) Regulations 2002
The Highways Act 1980
The Highways (Amendment) Act 1986
The Historic Buildings and Ancient Monuments Act 1953
The Housing Act 1957, 1961, 1964, 1969, 1974, 1980, 1985, 1988 and 1996
The Housing and Building Control Act 1984
The Housing Defects Act 1984
The Housing Grants, Construction and Regeneration Act 1996
The Insolvency Act 2000
The Land Registration Act 2002
The Landlord and Tenant Act 1954, 1985, 1987, 1988
The Landlord and Tenant (Covenants) Act 1995
The Latent Damage Act 1986
The Late Payment of Commercial Debts (Interest) Act 1998
The Law of Property (Miscellaneous Provisions) Act 1989
The Leasehold Reform, Housing and Urban Development Act 1993
The Licensing Act 1964, 1988, 2003
The Licensing (Amendment) Act 1980, 1981 and 1985, 1989
The Local Government Act 1999, 2000, 2003
The Local Government and Housing Act 1989
The Local Government Finance Act 1992
The Local Government (Miscellaneous Provisions) Act 1976 and 1982
The Local Government Planning and Land Act 1980
The Local Government Act 1963, 1972, 1985 and 1992
The London Building Act 1930
The London Building (Amendment) Act 1935, 1939
The Management of Health and Safety at Work Regulations 1999
The Management of Health and Safety at Work and Fire Precautions (Workplace) (Amendment) Regulations 2003
The National Heritage Act 1980 and 1983

Legislation

The National Heritage Act 2002
The National Lottery Act 1998
The New Roads & Street Works Act 1991
The Offices Shops and Railway Premises Act 1963
The Party Wall etc. Act 1996
The Petroleum (Consolidation) Act 1928
The Planning & Compensation Act 1991
The Planning and Compulsory Purchase Act 2004
The Planning (Consequential Provisions) Act 1990
The Planning (Hazardous Substances) Act 1990
The Planning (Listed Buildings and Conservation Areas) Act 1990
The Planning (Listed Buildings and Conservation Areas) (England) (Amendment) Regulations 2003
The Planning (Listed Buildings and Conservation Areas) (Amendment) (England) Regulations 2004
The Private Places of Entertainment (Licensing) Act 1967
The Property Misdescriptions Act 1991
The Public Health Act 1936 and 1961
The Rights of Light Act 1959
The Safety of Sports Grounds Regulations 1987
The Scheme for Construction Contracts (England and Wales) Regulations 1998
The Special Educational Needs and Disability Act 2001
The Sustainable and Secure Buildings Act 2004
The Theatres Act 1968
The Town and Country Planning Act 1990
The Town & Country Planning (Environmental Assessment & Permitted Development) Regulations 1995
The Town & Country Planning (Environmental Assessment) (England & Wales) Regulations 1999
The Water Act 1945, 1948, 1973, 1981, 1983 and 1989, 2003
The Water Act 2003
The Warm Homes and Energy Conservation Act 2000
The Water Industry Act 1999
The Water Resources Act 1991
The Workplace (Health, Safety and Welfare) Regulations 1992

Northern Ireland

There are some 108 national Acts relating to the design and construction of buildings in Northern Ireland.

The Building (Prescribed Fees) Regulations (Northern Ireland) 1997
The Building Regulations (Northern Ireland) 2000
The Construction (Design & Management) Regulations (Northern Ireland) 1995
The Construction (Design & Management) (Amendment) Regulations (Northern Ireland) 2001
The Control of Asbestos at Work Regulations (Northern Ireland) 1988, 2003
The Control of Asbestos at Work (Amendment) Regulations (Northern Ireland) 2000
The Control of Lead at Work Regulations (Northern Ireland) 1998, 2003
The Control of Substances Hazardous to Health Regulations (Northern Ireland) 2000, 2003
The Control of Substances Hazardous to Health (Amendment)

Legislation

Regulations (Northern Ireland) 2003
The Defective Premises (Landlord's Liability) Act (Northern Ireland) 2001
Disability Discrimination (Providers of Services) (Adjustment of Premises) Regulations (Northern Ireland) 2003
The Disability Discrimination Act 1995 (Amendment) Regulations (Northern Ireland) 2004
The Factories Act (Northern Ireland) 1965
The Fire Precautions (Workplace) Regulations (Northern Ireland) 2001
The Gas Safety (Installation & Use) Regulations (Northern Ireland) 1997, 2004
The Gas Safety (Installation & Use) Regulations (Northern Ireland) 2004
The Health & Safety (Safety Signs & Signals) Regulations (Northern Ireland) 1996, 2003
The Management of Health & Safety at Work & Fire Precautions (Workplace) (Amendment) Regulations (Northern Ireland) 2003
The Office & Shop Premises Act (Northern Ireland) 1966
The Planning (Compensation, etc.) Act (Northern Ireland) 2001
The Planning (Hazardous Substances) (Northern Ireland) Act 1993
The Pollution Prevention and Control Regulations (Northern Ireland) 2003
The Scheme for Construction Contracts (Northern Ireland) 1999
The Workplace (Health, Safety & Welfare) Regulations (Northern Ireland) 1993

Scotland

There are some 120 national and 50 Scottish Acts relating to the design and construction of buildings in Scotland.

NB: amending Acts (those shown with certain Acts) have important effects and should be read with them.

The Betting, Gaming and Lotteries Act 1963
The Building (Scotland) Act 1959 and 1970 as amended by the Housing (Scotland) Act 1986, 2003
The Building (Scotland) (Amendment) Regulations 1997
The Building (Scotland) Regulations 2004
The Building Standards (Scotland) Regulations 1990 (inc. amendments 1993-99)
The Building Standards (Scotland) Amendment Regulations 2001
The Building Standards (Scotland) Amendment Regulations 2001 Amendment Regulations 2002
The Building Standards and Procedure Amendment (Scotland) Regulations 1999
The Building (Procedure) (Scotland) Regulations 2004
The Celluloid and Cinematograph Film Act 1922
The Chronically Sick and Disabled Persons Act 1970 as extended by the Chronically Sick and Disabled Persons (Scotland) Act 1972 as amended by the Chronically Sick and Disabled Persons (Amendment) Act 1976
The Cinemas Act 1985
The Civic Government (Scotland) Act 1982
The Clean Air Acts 1956, 1968 and 1993
The Contaminated Land (Scotland) Regulations 2000
The Control of Pollution Act 1974
The Control of Pollution (Amendment) Act 1989
The Electricity (Scotland) Act 1979
The Factories Act 1961
The Fire Precautions Act 1971
The Food Safety Act 1990

Legislation

The Gaming Act 1968
The Gas Act 1972, 1986 and 1995
The Health & Safety at Work Act 1974 as amended by the Building Act 1984
The Housing (Scotland) Act 1987, 1988, 2001
The Land Reform (Scotland) Act 2003
The Late Payment of Commercial Debts (Scotland) Regulations 2002
The Licensing (Scotland) Act 1976
The Licensing (Amendment) (Scotland) Act 1992
The Local Government etc. (Scotland) Act 1994
The Local Government in Scotland Act 2003
The Local Government and Planning (Scotland) Act 1982
The Local Government (Miscellaneous Provisions) (Scotland) Act 1981
The Local Government and Housing Act 1989
The Local Government (Scotland) Act 1973, 1975, 1978 and 1994
The National Heritage (Scotland) Act 1985
The Offices, Shops and Railway Premises Act 1963
The Planning & Compensation (Scotland) Act 1991
The Planning (Consequential Provisions) (Scotland) Act 1997
The Planning (Hazardous Substances) (Scotland) Act 1997
The Planning (Listed Buildings & Conservation Areas) (Scotland) Act 1997
The Pollution Prevention & Control (Scotland) Regulations 2000
The Pollution Prevention and Control (Scotland) Amendment Regulations 2004
The Roads (Scotland) Act 1984
The Safety of Sports Grounds Regulations 1987
The Scheme for Construction Contracts (Scotland) Regulations 1998
The Scotland Act 1998
The Sewerage (Scotland) Act 1968
The Theatres Act 1968
The Town and Country Planning (Scotland) Act 1997
The Water (Scotland) Act 1980
The Water Industry (Scotland) Act 2002
The Water Environmental and Water Services (Scotland) Act 2003

Statutory instruments and orders

Many Acts of parliament empower secretaries of state and ministers to publish Statutory Instruments and Orders to implement legislation which, although on the Statute Book, require 'enactment'. The Building Regulations are a prime example. In these cases guidance is given as to how acts ought to be interpreted.

The Fire Precautions Act 1971

Under the 1971 Act, specifically designated classes of premises are required to have a fire certificate issued by the Fire Authority. Currently the Act applies to:

Hotels and boarding houses – except where no more than six people (staff and/or guests) have sleeping accommodation all on ground and/or first floors

Factories, offices, shops and railway premises – except where not more than 20 people work at any one time, or not more than ten persons work on any floor other than the ground floor, and shops and

Legislation

offices where only the employer's near relatives work or where not more than 21 hours weekly are worked.

In dealing with applications, the Fire Authority has to ensure that the means of escape from fire is adequate and effective and that the facilities for fighting fire and/or of giving warning in the event of fire achieve 'reasonable standards'. The existing or proposed use of the premises will be relevant.

If the standard is satisfactory a fire certificate must be issued but if it is not the Fire Authority must state the improvements required before they are prepared to issue a certificate.

A certificate will specify:

- ❖ the particular use or uses of the premises which the certificate covers;
- ❖ the means of escape in case of fire, including fire doors and fire escapes;
- ❖ the measures taken for securing safe and effective means of escape; and
- ❖ the type, number and location of the fire fighting equipment and fire alarms or warning systems in the premises.

A certificate can impose requirements ensuring that escape routes are kept clear, equipment is properly maintained and employees are adequately trained in fire drills. Further requirements may limit the number of people able to occupy the premises at any one time and can relate to other factors necessary to reduce the risk to persons in case of fire. The requirements can be imposed on the whole, or parts of the premises and may apply different requirements to the various sections. If the Fire Authority's requirements regarding the issue of a certificate seem unreasonable, the applicant may make appeal to a magistrate's court within 20 days of receipt.

Even if the premises are exempt under the Act or have been granted exemption by the Fire Authority, the Fire Precautions Act imposes a duty on the occupier and / or owner to ensure that reasonable provision is made for means of escape and means of fighting fire. There are penalties for contravening this duty.

The effect of the Fire Precautions (Workplace) Regulations 1997 and (Amendment) Regulations 1999 must be remembered. These regulations are described more fully below, but essentially supplement the existing provisions of the Fire Precautions Act. The 1971 Act still stands with the certification regime in force. However, there is now an additional responsibility placed on all employers to ensure that their premises have adequate fire precautions in place. This covers a wider range of premises than those covered in the 1971 Act, there now being only a very small number of workplaces that are exempt.

Fire Precautions (Workplace) Regulations 1997 and (Amendment) Regulations 1999

Following the introduction of the Fire Precautions (Workplace) Regulations on 1 December 1997 the Fire Precautions (Amendment) Regulations 1999 have been in force from 1 December 1999. These are in response to criticism from the European Commission on the failure of the earlier regulations to comply with the original directives. They exempted many premises that should have been covered and failed to reflect fully the unconditional nature of employers' responsibilities, namely that it ultimately rests with the employer, or the person responsible for the building (for example, the landlord), to ensure compliance with the regulations.

Legislation

Employer's duties

The 1997 regulations introduced fire to the realm of the Management of Health & Safety at Work Regulations 1992 (MHSW). These require employers to treat fire in the same way as any other health and safety issue and ensure that it is specifically covered in the risk assessment and in the comprehensive duties that follow the assessment. These include the provision of preventative protective measures, the use of competent persons, information provided to employees, incorporation and coordination with other employers and procedures for workers to follow where there is serious or imminent danger.

When the 1997 regulations were introduced it was the general assumption within the industry, and the Home Office's opinion, that those buildings covered by the UK system of fire certification would adequately comply. It has now been made clear that this is not the case and hence the new (amendment) regulations bring all work places under the regulations.

The general requirements are as follows:-

- ❖ Provide adequate fire fighting equipment, fire detectors and alarms. Any non-automatic fire fighting equipment such as extinguishers must be accessible, simple to use and indicated by signs. Employers must also take measures for fire fighting: nominating employees to implement those measures; ensuring they are trained and fully equipped; and making any necessary contact with the emergency services.
- ❖ Keep emergency routes and exits clear. Routes should lead to a place of safety as directly as possible, be indicated by signs and provided with emergency lighting. It must also be possible to evacuate the workplace as quickly and safely as possible, with emergency doors opening in the direction of escape.
- ❖ Maintain the equipment, routes and exits in sufficient state, working order and good repair.

It is important to stress that having a current fire certificate in place is not regarded as adequate demonstration of complying with the regulations. The emphasis has been very heavily placed upon the employer or the landlord, not only to ensure, but be able to demonstrate, that the appropriate risk assessments have been undertaken and appropriate action taken upon the findings of those risk assessments.

Enforcement of these regulations will still be carried out by the Fire Authority which will have the power to issue notices of non-compliance stipulating those works that must be undertaken within a set period of time. Failure to comply with the regulations or the requirements, or an enforcement notice will be a criminal offence.

The Fire Safety and Safety of Places of Sport Act 1987

This Act has made minor technical amendments to the FPA in the areas of fire certificate exemption; charges for the certification works; means of escape and fire fighting; interim duties as to the safety of premises; premises involving serious risk to persons; inspection of premises and other miscellaneous minor changes.

The remainder of this Act concerns general safety with amendments to the Safety of Sports Grounds Act 1975, primarily extending the scope of the Act with similar provisions for the safety of stands at sports grounds. The Act also includes fire schedules relating to the above.

The Disability Discrimination Act 1995

The purpose of the Disability Discrimination Act is to prevent discrimination against disabled people. Disabled people should not be treated less favourably in connection with employment, the provision of goods, facilities or services to the public, by those selling, letting or managing premises, education providers, trade organisations and qualifications bodies.

The Act is split into different sections.
Part 1 - defines the term 'disability'.
Part 2 - deals with discrimination in employment, trade organisations and qualifications bodies.
Part 3 - deals with discrimination in the provision of goods, facilities and services and the disposal and management of property.
Part 4 - deals with provisions for education.
Part 5 - deals with provisions for public transport.
Part 6 - deals with the setting up of the National Disability Council.
Part 7 - supplemental provisions.
Part 8 - miscellaneous provisions.

Part 1 - Definition of disability within the Act

A physical or mental impairment, which has a substantial and long-term (12 months minimum) adverse effect on a person's ability to carry out normal day to day activities. The proposed Disability Discrimination Bill proposes to extend the definition of disability in so far as to include more people with HIV, cancer and multiple sclerosis from the point of diagnosis. The main provisions of the draft Bill are provisionally timetabled for implementation from December 2005.

Legislation

Part 2 - Employment, trade organisations and qualifications bodies

Employment

This section of the Act came into force on 2 December 1996 and comes into effect when a disabled person is employed, or an employee becomes disabled. The Act places a duty on all employers, (the previous threshold exemption of less than 15 employees being removed from 1 October 2004) to provide accessible facilities for all disabled employees and to take all reasonable measures to enable disabled employees to carry out their work. The armed forces are excluded under the employment provisions.

An employer is expected to take reasonable measures to allow a person to do their job. This may involve:

- ❖ making adjustments to premises;
- ❖ moving a disabled person's place of work;
- ❖ altering hours of work;
- ❖ reallocating a disabled person's duties;
- ❖ acquiring or modifying equipment; and
- ❖ providing a reader or interpreter.

In addition to the above, employers should ensure that their equal opportunities policy addresses disability, in addition to preparing a disability statement and policy. Employers are not required to make

Legislation

changes in anticipation of employing a disabled person, however the code of practice suggests that employers should take opportunities to make improvements as they arise, eg as part of planned maintenance or refurbishment works. In addition, employers must not unjustifiably discriminate against current employees or job applicants on the grounds of disability, and may have to make reasonable adjustments to their employment arrangements or premises if these substantially disadvantage a disabled person.

Trade organisations and qualifications bodies

Qualifications bodies is an alteration to the trade association provisions of the DDA. These have come about as a result of a European Directive on discrimination in employment, which includes disability. These changes are contained in the Disability Discrimination Act 1995 (Amendment) Regulations 2003, which were implemented in October 2004.

The duty of a trade organisation or qualification body to make reasonable adjustments applies in respect of its disabled members and, also, in respect of any disabled person who is, or has notified the organisation that he may be, an applicant for membership.

Part 3 - The provision of goods, facilities and services including discrimination in relation to premises

The Act covers:

- any place where the public may enter;
- accommodation in hotels, boarding houses etc;
- all retailers or tradesman's premises;
- any building owned by a public authority; and
- facilities for entertainment or recreation.

Discrimination is deemed to have arisen if a provider of services treats a disabled person less favourably than he/she would treat others and:

- he/she cannot show the treatment in question was justified; and
- he/she failed to undertake his/her duty to make adjustments as described below.

Part III came into force in three distinct sections.

The following duties came into force in December 1996:-

- A duty not to discriminate.
- A duty not to refuse service.
- A duty not to provide a worse standard of service.
- A duty not to offer worse terms.

The following duties came into force in October 1999:-

- A duty to change practices, policies and procedures that are discriminatory.
- A duty to provide extra help such as auxiliary aids.
- A duty to overcome physical restrictions by the provision of alternative methods.

The following duties came into force in October 2004:-

Where a physical barrier makes use of any service which is offered to the public impossible or unreasonably difficult, a service provider must take reasonable steps to:

- remove the feature;
- alter it so it no longer has an effect;

- provide a reasonable means of avoiding the feature; or
- provide a reasonable alternative method of making the service available to disabled people.

The Act does not specify what is reasonable, as it varies according to the characteristics of each area of provision. Account may be taken of:

- the type of services being provided;
- the nature of the service provider and its size and resources;
- the effect of the disability on the individual disabled person;
- the amount of time that the service provider has had prior to that date to make preparations;
- whether taking any particular steps would be effective in overcoming the difficulty that disabled people face in accessing the services in question;
- the extent to which it is practicable for the service provider to take the steps;
- the financial and other costs of making the adjustment;
- the extent of any disruption which taking the steps would cause;
- the amount of any resources already spent on making adjustments; and
- the availability of financial or other assistance.

The Act does not require a service provider to take any steps which would fundamentally alter the nature of its service, trade, profession or business. Further interpretation of "reasonableness" is expected through the courts. The latest update on this can be found at the Disability Rights Commission web site: www.drc-gb.org.

Landlord responsibilities of common parts under the DDA

The duties of landlords under the DDA 1995 are unclear. The Code of Practice (Part 3) states that "there is no legal duty to make reasonable adjustments to premises which are sold, let or managed". In addition, "those who are selling, letting or managing premises do not have to make adjustments to make those premises more suitable for disabled people". The issue surrounding landlords responsibilities is complex and therefore landlords of premises with more than one occupier should not assume that they are not service providers for the purposes of the Act. The code of practice states that "they should anticipate that they may have responsibilities to make the common parts accessible to disabled people. They are advised to keep up to date with how the law in this respect is being interpreted". The code of practice goes on to say "if tenants are providing services to the public in their own right from the premises, they will have a duty under the Act to take reasonable steps to make the services accessible to disabled people.

Disability Discrimination (Providers of Services) (Adjustment of Premises) Regulations 2001

Applies to service providers and landlords of premises occupied by service holders.

The regulations prescribe particular circumstances in which it is reasonable or unreasonable for service providers to make physical alterations to the premises they rent or own or for lessors to withold their approval to same.

The provisions of the original DDA 1995 have been extensively amended and extreme caution should be taken when referring to material that may now be out of date. There is no substitute for professional legal advice in an environment where incorrect interpretation of the provisions is widespread.

Part 4 - Education

This part of the Act was amended by the introduction of the Special Educational Needs and Disability Act (SENDA) 2001

Part 1 makes changes to the existing legislation, in Part 4 of the Education Act 1996 (EA), for children with special educational needs.

Part 2 of the SENDA amends Part 4 of the Disability Discrimination Act 1995 (DDA) to introduce rights for disabled people in education.

Legislation

Publicly-funded providers of education services and private schools were exempt from Part 3 of the DDA. These provisions are repealed by this Act and the exemption removed. The effect of this is that any provider of education previously exempt from Part 3 and not covered by the new Part 4 duties becomes subject to the duties in Part 3 of the DDA.

The DRC have published two codes of practice that set out in detail the duties and responsibilities separately for both schools and post 16 education. The codes warrant detailed examination for those who are involved in educational establishments as they provide clarification and examples of potential scenarios covering the various duties and responsibilities under the Act.

The codes are available on the DRC website (see 'Further information' below) and are titled:

- ❖ Code of Practice for schools - DDA 1995: Part IV
- ❖ Code of Practice for providers of Post 16 education and related services - DDA 1995 - Part IV.

There are two key duties involved in ensuring that educational establishments do not discriminate against disabled pupils. These are:

- ❖ not to treat disabled pupils less favourably; and
- ❖ to take reasonable steps to avoid putting disabled pupils at a substantial disadvantage. This is known as the reasonable adjustments duty.

In framing the circumstances in which discrimination might occur, three broad categories are addressed:

- ❖ admissions;
- ❖ the provision of education and other activities or services; and
- ❖ exclusions.

Construction professionals should note particularly the differing requirements between schools and post 16 education. Duties under post 16 education applies to higher education, further education, adult and community education, schools providing further education for adults (but excluding sixth form) and youth and community services.

Part 5 - Public transport

Part 5 of the Act allows the government to set access standards for buses, coaches, trains, trams and taxis. Regulations have been introduced by the government to apply minimum access standards. Part 5 of the DDA applies specifically to the actual means of transport and does not include for example railway stations, ferry terminals, bus stations, airports and the like. This is covered under Part 3 of the Act. Under the code of practice, the following example is given; "A wheelchair user has no protection under Part 3 of the Act if a ferry on which he wishes to travel is not accessible. However, if he is refused service in the buffet bar of the ferry terminal because of his disability, this is likely to be unlawful."

Legislation

Codes of practice and advice notes

The government has drawn up various codes of practice to help implement the Act and interpret its requirements. Copies of the codes can be downloaded from the Disability Rights Commission website at www.drc-gb.org.

Further sources of information

Further information can be found on the website of the Disability Rights Commission at www.drc-gb.org.

DDA 1995 (Amendment) Regulations 2003

These regulations generally came into force on 1 October 2004 although they already allow the Disability Rights Commission (DRC) to prepare new codes of practice (see above). The regulations extensively amend the Part 2 (Employment) duties of the DDA and other legislation and are aimed at implementing the provisions of European Directive 2000/78/EC relating to employment etc. Key issues relating to employers include inter alia:

- ❖ Removal of 15 employee exemption (although excluding the armed forces).
- ❖ Extension of Part 2 provisions to those who may not strictly be defined as employees (such as contract workers etc) and qualifying organisations.
- ❖ Removal of exemption for employers to make reasonable physical adjustments in respect of buildings where elements meet the requirements of Part M of the Building Regulations or Scottish equivalent. The exemption for service providers under Part 3 of the DDA is unchanged.
- ❖ Inclusion of two additional factors to be taken into account on the question of reasonableness under Part 2 of the Act.

BS 8300:2001 'Design of buildings and their approaches to meet the needs of disabled people - Code of Practice'

This is an important document in the field of access for disabled people and provides extensive and detailed guidance on good practice in the design of and access to buildings and their approaches. It forms the basis for much of the amendment to the Part M Approved Document under the Building Regulations. This document is much more comprehensive than the Approved Document and therefore, it may not be sufficient to rely uniquely on Part M.

Part M Approved Document

This recently revised guidance is based on and is complementary to BS 8300:2001, although the BS contains much additional material that is not included in the new Part M Approved Document. In some cases, the guidance in the Part M Approved Document differs from the recommendations in BS 8300. Compliance with the recommendations in the BS, therefore, while ensuring good practice, is not necessarily equivalent to compliance with the guidance in the Part M Approved Document and appropriate care should be taken when dealing with both documents.

Disability Discrimination Bill

A new Disability Discrimination Bill is expected which is provisionally timetabled to be introduced from December 2005. This was published in draft form in December 2003. Full details can be accessed at the following web address; www.disability.gov.uk. The proposed Bill will amend the current Disability Discrimination Act 1995 in some of the following ways:

- ❖ Definition of disability - extended so far as to include more people with HIV, cancer and multiple sclerosis from the point of diagnosis.
- ❖ Transport - the DDA is to be extended to cover discrimination in relation to transport.
- ❖ Public sector - Promote disability equality.

Legislation

- Public authorities - the DDA will be extended to cover most functions of public authorities.
- Letting of premises - the DDA duties will be extended to include a duty to make reasonable adjustments to policies, practices and procedures and to provide auxiliary aids and services, where it is reasonable to do so. Landlords will therefore have a duty to take reasonable steps to facilitate access for disabled tenants. However, landlords will not be under any duty to carry out adjustments to physical features of the premises. Furthermore, the cost of any reasonable adjustments that a landlord may have to make (in respect of policies, practices, procedures, auxiliary aids and services), cannot be recovered by way of increased rent or service charges.
- Private clubs - any club with 25 or more members will be covered by the DDA
- Third party publishers (eg newspapers) would be liable for publishing discriminatory advertisements, so that they would be acting unlawfully if they published a discriminatory job advertisement.
- Insurance provided on group terms to an employer's staff is to be covered as a service under the DDA.

The National Register of Access Consultants

The National Register of Access Consultants (NRAC) was established in 1999 to accredit access auditors and access consultants. This is the only recognised UK body of its kind. Its web site can be accessed at www.nrac.org.uk. There are two levels of membership; auditor or consultant. The fundamental difference between auditor and consultant is that consultants are required to have construction-related qualifications. Therefore, there are some limitations to the extent and detail of the advice that auditors can provide. Construction professionals such as surveyors and architects as qualified consultants are well placed to identify access issues, provide technical recommendations and also indicative costs.

Disability facts

- There are over 8.5 million disabled people in the UK, all of whom are potential customers.
- It is estimated that disabled people spend around £40 billion a year on goods and services.
- There are more than two million disabled people in employment in the UK.
- There are at least another million disabled people who want jobs but are out of work – many of whom are equally skilled as those who have jobs.
- Disability is too often associated with wheelchairs but only 5% of disabled people use wheelchairs.

Building and construction regulations

➡ The Building (Amendment) Regulations 2004 — 58

➡ Airtightness — 58

➡ Health and safety at work for surveyors — 59

➡ The Construction (Health, Safety and Welfare) Regulations 1996 — 61

➡ The Construction (Design and Management) Regulations 1994 — 63

➡ The Workplace (Health, Safety and Welfare) Regulations 1992 - Regulation 14 — 65

➡ Workplace Regulations - provision of sanitary facilities — 66

Legislation

The Building (Amendment) Regulations 2004

Building Regulations state-of-play table

Part A Structure	New Regulations Approved Document A - Structure (2004 Edition) In effect from 1 December 2004.
Part B Fire safety	Came into effect on 1 March 2003 2000 edition consolidated with 2000 and 2002 amendments.
Part C Site preparation	New Regulations made 28 May 2004. In effect from 1 December 2004.
Part E Resistance to the passage of sound	Amended 2004 to include clarifications and correction of errors in the 2003 edition.
Part L Conservation of fuel and power	Substantial changes expected around 2008
Part M Access to and use of buildings	2004 edition published
Part P Electrical safety	Comes into effect 1 January 2005 Excludes work carried out by an approved competent person and work of a minor nature

Airtightness

Parts L1 and L2 of the Building Regulations contain, among other things, a requirement for minimising air leakage from buildings.

Achieving an airtight building means following three essential steps:-

- ❖ Design for airtightness.
- ❖ Build for airtightness.
- ❖ Test for airtightness.

It is not practicable to construct a building and then try to make it airtight. Remedial sealing can be difficult and costly. By designing in airtightness at the drawing stage you can deal with air barrier continuity and sealing details at critical elements – and ensure long-term performance by specifying the correct seal or sealant.

The main air leakage problems in buildings occur typically:

- ❖ around doors, windows, panels and cladding details;
- ❖ in gaps where the structure penetrates the construction envelope;
- ❖ in service entries: pipes, ducts flues, ventilators;
- ❖ in porous construction: bricks, blocks, mortar joints; and
- ❖ in joist connections within intermediate floors.

Designers should identify all the problem areas, for example, sealing around pipe entries, and spell out responsibility for finishing off in the contract documents.

Legislation

Constructing the building to the airtightness specification is then down to the main contractor and sub-contractors. For this to be successful all of the work force should be aware of airtightness issues in the same way safety issues and codes of conduct are dealt with.

Inspection during construction is essential. Talking to and working with contractors is the best way of ensuring that the team understands the importance of the airtightness layer and how to incorporate it.

The only real way to be confident that the building meets an airtightness specification is to carry out a fan pressurisation test prior to handover as required in the proposed revisions to Part L. Large buildings, for example, hypermarkets or industrial buildings, need specialist larger capacity equipment.

If the three essential steps listed above are followed, the building should pass the test. In the event they are not and the building fails, the proposals in Part L state: "If on first testing the building fails to comply, the major sources of air leakage should be identified using the techniques described in TM23 [CIBSE Technical Memorandum]." This usually requires specialist help.

Despite improved understanding of construction techniques, few buildings are sufficiently airtight – this is true of new buildings as well as old. In a recent survey, only one out of 39 buildings tested met a good practice benchmark for airtightness. While the degree of leakage varies considerably from building to building, it is not unusual for the problem to be equivalent to having a $9m^2$ hole in the building envelope.

Both the government and CIBSE regard airtightness as a serious issue and encourage protective measures.

Airtightness testing

CIBSE has produced a guide titled 'Testing Buildings for Air Leakage' (TM23: 2000).

The technique that is most commonly used to measure air leakage is 'fan pressurisation testing'. This involves a specially designed system of fan units that blow air into the building, and the measurement of air leakage from the building at various air pressures.

Testing can be carried out on any building from a large hypermarket, multi-storey office block or factory to a small store, office or even an individual room within a building. Testing a large office or superstore takes about three hours and is generally done out of business hours when the premises are closed.

Health and safety at work for surveyors

Essential legal duties

The Health & Safety at Work etc. Act 1974 requires all chartered surveyors to ensure, so far as is reasonably practicable, the health and safety of themselves and any other people who may be affected by their work.

In essence, the places where surveyors' work must be safe and working practices must be clearly defined, organised, and followed to avoid danger. This requires safety training and the distribution of relevant information, followed up by diligent and regular supervision.

These duties extend to anyone who uses surveyors' professional services.

Those who lease part of their premises to other businesses also may be responsible for them with regard to safety matters.

In addition, all working people whether employees, managers, partners or directors (self-employed or not) must behave in a way that does not endanger themselves or others.

Legislation

Health and safety policy statement

Every firm of surveyors which employs five or more people is obliged by the Act to draw up a health and safety policy statement, which should be kept up to date, with any significant revisions being notified to employees.

The Health & Safety Executive (HSE) in "Writing your health & safety policy statement: how to prepare a safety statement for smaller businesses" describes what the document should say and gives a useful pro-forma.

It is suggested that employees, especially when taking a new job, should satisfy themselves that they have seen and understood the company's policy statement and have been made familiar with safe working practices. If they consider that the Health & Safety at Work Act is not being followed, they ought to say so, and if necessary ask advice from their local HSE office (listed in the telephone directory).

Practical procedures

Chartered surveyors should identify the hazards they may encounter in practice, carry out a risk assessment and plan accordingly. The way they proceed will depend upon the working environment: when surveying old and derelict buildings, for example, there may be holes in floors, parts of the structure may be unstable or there may be health hazards. Particular care should be taken to protect against personal attack.

The Control of Substances Hazardous to Health Regulations 2002

The purpose of The Control of Substances Hazardous to Health Regulations 2002 is to safeguard the health of people using or coming into contact with substances that are hazardous to health.

Substances are classified as being very toxic/toxic, harmful, corrosive or irritant. Under these regulations employers are required to evaluate the risk of all products used that may be harmful to the health of their employees and take appropriate measures to prevent or control exposure.

The 'Six Pack' Regulations

The original six sets of regulations, introduced in 1993, were wide ranging and, with some minor exceptions, apply to all places of work. The previous style of legislation was to apply individual acts or regulations to individual industries or sectors of industry, hence the Factories Act, the Offices, Shops and Railway Premises Act, and the Construction Regulations.

Workplace (Health, Safety and Welfare) Regulations 1992

These apply to all workplaces. The regulations set out the minimum requirements in respect of the provision and maintenance of the environmental conditions, space allocation, sanitary and welfare provision of employees. In addition, they impose particular safety requirements on forms of construction or circumstances which are considered to be high risk. They do not apply to construction sites.

Provision and Use of Work Equipment Regulations 1998

These apply to all workplaces. Basically, all existing and new work equipment which includes everything hired or purchased second hand, must comply with the regulations.

Every employer must ensure that all work equipment is so constructed or adapted as to be suitable for the purpose for which it is used or provided. They identify specific hazards that the employer must prevent or adequately control.

Legislation

Manual Handling Operations Regulations 1992

These require the employer to try to avoid the need for employees to undertake any manual handling operations at work that involve a risk of their being injured. Where this is not reasonably practicable the risk must be assessed and suitable provision made, including equipment, instruction and training for safe manual handling.

Management of Health & Safety at Work Regulations 1999

These regulations are of a wide-ranging general nature and overlap with many others. They require the employer to carry out an assessment of the risks of the hazards to which his/her employees are exposed at work or to others arising from or in connection with this work. He/she must instigate appropriate protective or preventive measures, reviewing and amending these as necessary.

The employer must appoint a 'competent person' to provide assistance in respect of these duties. Emergency procedures must be put in force to deal with any serious and imminent danger. Employees must be informed of these measures and suitably trained where required. They are obliged to comply with these instructions and warn of any situation considered to be a serious and immediate danger to health and safety.

Personal Protective Equipment at Work Regulations 1992

Under these regulations the employer has a duty to provide and maintain suitable personal protective equipment including adequate instruction and training on its correct use when risks to health and safety cannot be avoided by other means. Employees have a duty to make full and proper use of such equipment provided and to report any loss or obvious defects.

Display Screen Equipment Regulations 1992

The need for the regulations is primarily the evidence of repetitive strain injury that is reported by keyboard users, the amount of time lost due to other causes of sickness among users and, of course, the European Directive.

These regulations target full time users of visual display units, mainly in the banking, insurance and data processing sectors but, given that most of the medium and larger companies will have in their offices a number of full time or habitual users, then these regulations will apply. They will also apply in the office facility of a construction site, if any persons are habitual users of visual display units.

Construction (Health, Safety and Welfare) Regulations 1996

The CHSW Regulations came into force on 2 September 1996, to form a single set of regulations applicable to construction work and construction sites. They consolidate, modernise and simplify much of the previous legislation and complete the EC directive on construction.

Their aim is to protect the health, safety and welfare of everyone "carrying out construction work" and also to protect those affected by the work.

The regulations impose requirements with respect to:

- ❖ the provision of safe places of work and safe access and egress thereto (regulation 5);
- ❖ the provision of suitable equipment to prevent falls (regulation 6);

Legislation

- working on or near fragile material (regulation 7);
- the prevention of injury from falling objects (regulation 8);
- the stability of structures (regulation 9);
- the carrying out and supervision of demolition and dismantling and the use of explosives (regulation 10, 11);
- the safety of excavations, cofferdams and caissons (regulation 12, 13);
- the prevention of drowning (regulation 14);
- the movement of pedestrian and vehicular traffic (regulation15);
- the construction of doors, gates and hatches (regulation 16);
- the use of vehicles (regulation 17);
- the risks from fire, the provision of emergency routes and exits, the preparation and implementation of evacuation procedures and the provision of fire-fighting equipment, fire detectors and alarms (regulations 18, 19, 20, 21);
- the provision of sanitary and washing facilities, a supply of drinking water, rest facilities and facilities to change and store clothing (regulation 22);
- the provision of adequate fresh air, reasonable temperature and weather protection (regulations 23, 24);
- the provision of lighting, including emergency lighting (regulation 25);
- the marking and good order of a construction site (regulation 26);
- the safety and maintenance of plant and equipment (regulation 27); and
- training and supervision (regulation 28).

Points to note (including changes of current practice)

Application of the regulations

The Health and Safety Executive is, in certain circumstances, able to issue exemption certificates subject to conditions or time limitations deemed appropriate and which may be subsequently revoked if considered necessary.

Subjective requirements

As with other recent health and safety legislation, the regulations have moved away from prescriptive demands and most of the requirements are, to the extent that they are 'suitable and sufficient', to be determined on the basis of a risk assessment of the particular circumstances. In addition, many are qualified by their 'reasonable practicability', which means that the action to be taken should be proportionate to the risk involved.

Scaffolding

This is one instance where the requirements are prescriptive and the measurements stated mean that virtually all scaffolding and working platforms will require an intermediate guard rail or other means of physical protection, such as a brick guard.

Welfare

The responsibility for the provision of suitable and sufficient welfare facilities is now that of the "person in control of a construction site" as well as the employer of the site operatives or the self employed.

The washing facilities are to be appropriate to the nature of the work and there is an inference that showers may be required in certain circumstances.

Legislation

The Health and Safety Executive is paying great attention to the adequacy of welfare facilities on sites and has pointed out that some wash hand basins will need to be of sufficient depth to enable the complete forearm to be immersed for proper cleaning. On occasion, basins have been kept deliberately shallow to minimise the water required.

Falls

Particular emphasis is put on the measures to be taken to prevent persons falling from a height and the regulations set out a hierarchy of alternative methods by which this may be achieved. This takes into account practicability, physical constraints and the duration of the work. The use of safety harnesses, as with all personal protective equipment, should be a last resort.

Prevention and control of emergencies

There are specific requirements to plan for unforeseen circumstances that may arise on construction sites. To prevent risk from fire, explosion, flooding and asphyxiation, procedures for evacuating the site must be established, including emergency escape routes and, where necessary, fire fighting equipment, fire detectors and alarm systems must be provided.

Traffic movement

There are requirements to ensure the safe use and movement of vehicles used in connection with construction work and the provision of safe traffic routes generally.

Further information

For further details, reference should be made to the regulations themselves (ISBN 0-11-035904-6 available from HMSO).

In addition, the HSE has produced a guidance publication 'Health & Safety in Construction' which describes practical ways of complying with the regulations. (ISBN 0-7176-1143-4).

The Construction (Design and Management) Regulations 1994

The Construction (Design and Management) Regulations (CDM) 1994 provide the framework for managing health and safety during the construction, repair, maintenance and demolition of civil engineering and building works.

They impose statutory duties on clients, designers and planning supervisors (a new role) and contractors.

The principal objectives are to:

- ❖ ensure proper consideration and coordination of health and safety issues at every phase of a project, from feasibility study to demolition;
- ❖ obtain adequate allocation of resources (including sufficient time) to enable duties imposed by the regulations to be met;
- ❖ involve directly, allocate and share responsibilities between all participants (including the client); and
- ❖ promote the appointment of competent (from a health and safety viewpoint) designers, planning supervisors and contractors.

To achieve its aims, for each new project to which the regulations apply, two documents are required:-

Health and safety plan – in two stages, planning and construction, to convey health and safety information to the contractor during the tender and construction phases of the project.

Legislation

Health and safety file – a record of information relevant to health and safety to be retained by the client, made available and used throughout the life of the building or structure to assist its safe maintenance, alteration or demolition.

The client must appoint a health and safety coordinator. This role is separated for the pre-contract planning and post-contract implementation phases of the project: ie the planning supervisor, who is ultimately responsible for the preparation and coordination of the safety plan, up to the appointment of the principal contractor. The principal contractor takes over responsibility for developing and implementing the safety plan during the construction phase of the project.

Construction Regulations 1994

Revised Approved Code of Practice 'Managing Health & Safety in Construction' published by Health & Safety Commission (HSG 224)

The new ACOP came into effect on 1 February 2002.

Its aim is to rectify confusion and shortcomings identified in the previous publication, in particular to improve the focus of work carried out by the various dutyholders and reduce bureaucracy.

The ACOP has been re-formatted in a more logical, user-friendly layout and emphasises the financial and other benefits of the CDM regulations, provided they are implemented appropriately.

It clarifies a number of matters which were previously often misinterpreted, by listing those matters which are not required to be included in health and safety documents or which are excluded from the responsibility of particular dutyholders, by giving examples of both good and bad practice and by changing the emphasis of particular issues.

The topics where there are significant changes are summarised below and, for each, the concern to be addressed is described (in italicised text) followed by the suggested means of improvement.

Competence and resources

Expense and wasted effort arising from use of assessment questionnaires and processes that are often excessive, unnecessary and inappropriate.

Assessments should: form part of the more general checks routinely carried out (for example, quality, finance and viability), focus on the particular project and be proportionate to the risks, size and complexity of the work.

Time and programme

Dutyholders not appointed early enough in the project, design considerations made too late and (design and construction) programmes unrealistic.

- ❖ Planning supervisor, principal contractor and other key people must be appointed as soon as possible.
- ❖ Design risks to be identified and removed or reduced early in the design process, at the concept and scheme stages, not left until detail design.

Design and designers

Using unnecessarily sophisticated risk analysis techniques; identifying risks but not adapting the design to remove or reduce them and overlooking less obvious 'designers'.

- ❖ A team approach to risk consideration is recommended.

Legislation

- ❖ ACOP lists untypical designers, including temporary works engineers, interior designers, shopfitting, trade contractors and manufacturers of purpose-made products.
- ❖ ACOP lists, ways in which the designer can have direct positive influence on risks and circumstances in which relevant information must be provided to others together with those matters which are not the designer's responsibility.

Clients

Failing to appreciate the full scope of the definition of a 'client'; unnecessarily monitoring performance of others and not providing information, either at all or in sufficient time.

- ❖ ACOP gives examples of other parties who may be 'project originators' and thus take on 'client duties'.
- ❖ No legal duty under CDM regulations to review assessments, monitor performance of appointees or continued adequacy of the health and safety plan during the construction phase.
- ❖ ACOP provides examples of relevant information to be provided to the planning supervisor early enough for implications to be assessed by the designer.

Health and safety plan and file

Inclusion of unnecessary and irrelevant information that can affect detrimentally the ease of identification and significance of crucial details.

- ❖ Appendix 3 contains a list of matters that should be included or addressed within the health and safety plan, where relevant to the work proposed. This is in tabular format with those of the pre-tender stage alongside those of the construction phase under the same topic headings, to illustrate and compare the appropriate requirements for each.
- ❖ Appendix 4 has a similar list of contents of the health and safety file and the ACOP also lists those topics that do not need to be included.

Consultation of construction workers

Decisions on health and safety arrangements are often made without consulting workers.

ACOP extols the necessity for and the benefits of obtaining feedback from workers' formal or informal safety committees or representatives, by addressing common problems, reviewing accidents and near misses, and identifying and considering how risks should be addressed on site.

The Workplace (Health, Safety and Welfare) Regulations 1992 – Regulation 14

The regulations require that glazed doors (and gates) be fitted with safety materials where any part of the glazing is below shoulder height. This requirement applies to glazing in the panels at the sides of the doors (and gates) because these areas are often struck or pushed when mistaken for part of the door.

The requirements also apply to windows, walls and partitions where there is glazing below waist height.

In situations where the width of the glass panel exceeds 250mm then

Legislation

safety materials must be used. Safety materials include:

- ❖ polycarbonates, glass blocks or other materials that are inherently robust; or
- ❖ glass that will break safely (ie shatters without a chance of sharp edges) or ordinary annealed glass that is of sufficient thickness relative to its area, as outlined in the following table.

Nominal thickness	Maximum size
8mm	1100 x 1100mm
10mm	2250 x 2250mm
12mm	3000 x 4500mm
15mm	Any size

Clearly therefore, just because annealed glass exists, it does not follow that additional protection is required.

Glazing should always comply with British Standard BS 6262: 1994, Code Of Practice for Glazing in Buildings safety related to human impact.

In circumstances where glazing does not comply, it will be necessary to replace it with safety materials or to provide some physical protection that will ensure that it meets the impact performance levels required by BS 6262, Part 4: 1994.

A cheaper alternative would be to apply safety film to achieve the BS 6262 standard. Manufacturers of film should be consulted to ensure an appropriate grade of material for glass size.

All glass should be suitably marked as being of a safety standard. Where glass is protected by film, labelling should identify this.

Identification

Laminated and toughened glass can be detected using proprietary glass testing kits.

Note that ordinary Georgian wired glass does not comply with safety standards, but Georgian wired safety glass does.

Workplace Regulations - provision of sanitary facilities

The Workplace (Health, Safety and Welfare) Regulations require that sanitary provision shall be suitable and sufficient for the numbers and types of workers employed.

BS 6465: Part 1: 1994 sets out the minimum requirements.

It contains numerous tables illustrating the number and types of sanitary appliances required for a variety of different circumstances, depending on: the number of workers, and where relevant, customers; the ratio of males to females; and the different primary uses of the building in question.

The types of building include offices, shops, factories, restaurants, cafes, canteens and fast-food outlets, swimming pools, stadia, public houses, licensed bars, other places of public entertainment and non-domestic premises.

Legislation

Comparison of requirements for 100 people evenly divided between the sexes (not including wheelchair users)

Building type	Male			Female		Total		
	wc	urinal	whb	wc	whb	wc	urinal	whb
Workplaces	3	2	3	3	3	6	2	6
Workplaces (dirtier conditions)	3	2	3	3	5	6	2	10
Shops (customers) 1000 - 2000sqm	1	1	1	2	2	3	1	3
Shops (customers) 2000 - 4000sqm	1	2	2	4	4	5	2	6
Restaurants	1	1	2	2	2	3	1	4
Public houses etc	1	2	2	3	2	4	2	4

wc = water closet   whb = wash hand basin

Town and country planning in England

- Development applications and fees — 70
- Appeals and called-in applications — 70
- Development plans and monitoring — 70
- General Development Order and Use Classes Order — 71
- Planning policy guidance notes and circulars — 72
- Listed buildings and conservation areas — 73
- Environmental impact assessments — 75

Legislation

Legislation

Caveat
At the time of going to press, considerable changes are being introduced to the planning system. The Planning and Compulsory Purchase Act was passed in late 2004 and Commencement Orders are being introduced in phases. Guidance in the form of Planning Policy Statements now includes: PPS7 - Sustainable development in rural areas; PPS22 - Renewable energy; PPS11 - Regional planning; and PPS12 - Local development frameworks. The latter two herald a new shape for Development Plans. Planning obligations are likely to come under a stricter régime.

Development applications and fees

Building, engineering, mining or other operations in, on, over or under land, including demolition, all require development permission, but many minor matters have permitted development rights (General Permitted Development Order) and demolition consent is only required in specific cases (Circular 10/95). Material changes of use of land or buildings require permission but many have deemed permission (Use Classes Order and General Permitted Development Order).

Buildings which are listed as of special historic or architectural interest (Grades 1, 2* and 2) may not be altered (or demolished) in any way that might affect their character as a listed building without listed building consent. Unlisted buildings in a conservation area may not be wholly demolished without Conservation Area Consent. Scheduled monuments may not be altered at all without consent. Advertisements may not be erected without Advertisement Control Consent but many have deemed consent (Advertisement Control Regulations). Applications should be dealt with in eight weeks by the Local Planning Authority or in 16 weeks if an environmental statement is submitted. There is *no* statutory period of 13 weeks. However, the Government will be consulting on extending this period in 2005.

A fee is payable for every planning and advertisement application but the Application Fee Regulations allow for some exemptions or reductions normally related to the type of applicant, or where an earlier application has been submitted and refused or withdrawn. There are no fees for entering planning appeals but there are for enforcement appeals. From the 1st April 2005, there will be major increases in fees and the maximum fee will rise to £50,000.

Appeals and called-in applications

All types of application give rise to a right of appeal against refusal, non-determination or conditional approval. Appeals are dealt with by the Planning Inspectorate, an executive agency of central government reporting to the Office of the Deputy Prime Minister (ODPM). Appeals must be made within a certain time from the date of the decision of the relevant council (or within six months of when they should have made a decision), and can only be made on specific appeal forms available from the Planning Inspectorate.

There are three types of appeal procedure – public inquiry, hearing or written representations – secondary legislation, in force since 1 August 2000, governs the timetabling leading to the inquiry, hearing or site inspection, but there is no time limit on the decision-making process itself. As at November 2004 decision-making times at appeal are the longest ever recorded. DETR Circular 5/2000 summarises the appeal procedures.

Called-in applications are rare but have the effect of opening up the application to public debate at an inquiry, with the decision being made by the ODPM and not the Local Planning Authority.

Development plans and monitoring

Where the development plan is material to the development proposal and must therefore be taken into account, Section 54A of the 1990 Planning

Legislation

Act requires the application or appeal to be determined in accordance with the plan, unless material considerations indicate otherwise. In effect, this introduces a presumption in favour of development proposals, which are in accordance with the development plan (Planning Policy Guidance Note No.1: General Policy and Principles). Section 38 of the 2004 Act will take the place of s54A.

The primacy of development plans is now enshrined in the planning system and ignoring the opportunity to make representations on development plans can seriously damage development prospects and limit the opportunities available for redevelopment or changes of use. Similarly applicants, or their agents, must be aware of the relevant policies in the development plan before entering any application. The monitoring of development plans is therefore an essential element of both sound estate management and of preparation for development. The 2004 Act introduces a replacement of existing plan making procedures with Local Development Frameworks and Regional Spatial Strategies now to be prepared.

General Development Order and Use Classes Order

The Town and Country Planning (General Permitted Development) Order 1995 (amended 2001)

The GPDO gives a general permission for certain defined classes of development or use of land, mainly of a minor character. The most commonly used class permits a wide range of small extensions or alterations to dwelling houses, and there are many categories that allow for development that would otherwise require planning permission. There are allowances for developments which would also require listed building, conservation area or advertisement control consent, but this does not remove the need for such consents.

The general permission that the GPDO grants for a particular development or class of development may be withdrawn in a defined area by an Article 4 Direction made by the local authority or the ODPM.

The Town and Country Planning (General Development Procedure) Order 1995 (as amended)

This order came into force in June 1995, as did the GPDO, and details procedural matters on applications, including a full list of statutory consultees, as well as requirements for registers of applications, specimen copies of forms and guidance on appeal submissions.

The Town and Country Planning (Use Classes) Order 1987 (as amended)

Section 55 of the 1990 Planning Act provides that changing certain uses does not constitute development under the Act and this includes any change of use of land or buildings from one use to another within the same class of the Use Classes Order. Other changes of use are not necessarily development, only being so when the change is material; normally when changing from one use class to another outside the allowances in the UCO or GPDO, or when the new use is of a different character to the old use.

A 1993 change to the UCO removed hostels from the hotels and hostels use class (Class C1) and there is now much case law clarifying changes of use. Circular 11/95 of the DoE makes it clear that there is a presumption against planning conditions designed to restrict further changes of use which, by virtue of the UCO, would not otherwise constitute development.

Since the UCO was published in 1987, many organisations have produced

Legislation

comparative tables of permitted changes but in view of the refinement and updating of the UCO these are now suspect and each case should be considered in the light of the current legislation. The courts have held that it is not necessary to go to extreme lengths to identify a class for every use and the UCO lists a number which are not to be regarded as being in any class of the order. This does not necessarily mean that a change to, or from, such a use, to a use within a specific class will always be a material change of use requiring planning permission.

Uses which have an uninterrupted history of ten years, but which do not have planning permission, may now be the subject of an application for certification of the lawful use commensurate with a planning approval.

Planning policy guidance notes and circulars

As at November 2003, there were 25 planning policy guidance notes (PPGs) and planning policy statements (PPSs) as identified in the following table. These are to be taken into account by local authorities as they prepare their development plans, and they may be material to decisions on individual planning applications and appeals. Guidance in the PPGs and PPSs can be considered as particularly material if the development plan does not reflect the policy direction in them.

PPG/PPS Number	Title	Date
1	General policy and principles (r)	1997
2	Green belts	1995
3	Housing (r)	2000
4	Industrial/commercial development and small firms	1992
6	Town centres and retail developments (r)	1996*
7	Sustainable development in rural areas	2004
8	Telecommunications (r)	2001
9	Nature conservation (r)	1994
10	Planning and waste management (r)	1999
11	Regional spatial strategies	2004
12	Local Development Frameworks	2004
13	Transport	2001
14	Development on unstable land	1990
15	Planning and the historic environment (see circular 1/2001 for amendments)	1994*
16	Archaeology and planning	1990
17	Open space, sport and recreation	2002
18	Enforcing planning control	1991
19	Outdoor advertising control	1992
20	Coastal planning	1992
21	Tourism (r)	1992
22	Renewable energy	2004
23	Planning and pollution control	2004
24	Planning and noise	1994
25	Development and flood risk	2004

* = amended by ad hoc statements/court cases

(r) = under revision

Some PPGs have one or more 'annex' and/or Best Practice Guides.

Legislation

Since May 1996 The Welsh Office has prepared technical advice notes and planning policy guidance just for Wales. The Scottish Office publishes planning advice notes. Regional planning guidance is now contained within RPGs and mineral planning guidance in MPGs for England.

Circulars

Circulars are now used primarily to publicise technical and administrative changes but there are many that are still particularly relevant, notably:

- Circular 13/87: The Use Classes Order
- Circular 06/98: Affordable Housing
- Circular 5/94: Planning out crime
- Circular 9/95: GDO Consolidation
- Circular 10/95: Demolition
- Circular 11/95: Planning Conditions
- DETR Circular 2/99: Environmental Impact Assessment
- DETR Circular 5/2000: Planning appeals
- DETR Circular 1/2001: Heritage Applications
- GOL Circular 1/2000: Strategic Planning in London.

The ODPM has introduced a housing density circular concerning the south-east (1/2002).

Greater London

The Mayor of London is responsible for certain planning functions and has published regional planning guidance and strategies in the adopted London Plan (2004).

Listed buildings and conservation areas

'Planning and the historic environment' (PPG15) places great stress on keeping historic buildings in active use and urges planning authorities to be flexible and imaginative in their approach, to achieve the right balance between protecting a building's special character and adapting it for new uses. A new system trialing in 2004 seeks to record all historic designations on one property under a single 'listing'.

Listed buildings in England

The principal current legislation is the Planning (Listed Buildings and Conservation Areas) Act 1990. This sets out the statutory framework relating to listed buildings and conservation areas.

Listed Building or Conservation Area Consent is not required for a material change of use. Change of use is covered under the Planning Act 1990.

Grade 1:
Buildings of exceptional interest (about 2% of listed buildings).

Grade 2*:
Particularly important buildings of more than specialist interest (about 4% of listed buildings).

Grade 2:
Buildings of special interest which warrant every effort being made to preserve them.

Legislation

Listing

The basic criteria for listing are set out in PPG15 and while 'spot listing' will still be considered for individual buildings overlooked or under threat, by reference of details by any member of the public to the ODPM/English Heritage, future research to identify potential buildings for listing will alter. The list reviews will concentrate on finding the best examples of types and periods of buildings under-represented in the current list.

What is a listed building?

Any structure or erection, and any part of a building including any object or structure within the curtilage that forms part of the land and did so before 1 July 1948.

Material change of use

It is accepted that new uses for old buildings may often be the key to their preservation. In some instances, it may be appropriate that control should be relaxed where this would enable historic buildings to be given a new lease of life.

The best use for an historic building is obviously the use for which it was designed and, wherever possible, the original use should continue.

Works for which listed building consent is necessary

Demolition of a listed building, or its alteration or extension in any manner that would affect its character, requires consent. It should be noted that the setting of an historic building is considered of great importance and an essential feature of its character. Consent is not required for works of repair that are on the basis of like-for-like materials and exact details. However, this is a grey area.

Application for listed building consent

Applications for listed building consent should be made to the local planning authority.

Listed building consent decisions

The local planning authority has eight weeks in which to consider an application. This can be extended by agreement with the applicant or revision of proposals, etc.

The legislation requires all listed building consents to have conditions, even if only time/recording conditions.

Listed building consent for works already executed

Listed building consent may be sought even though the works have already been completed. If consent is granted this is not retrospective; the works are authorised only from the date of consent.

Conservation areas

Local authorities have a duty imposed by section 71 of the Listed Buildings Act 1990 that they must regularly formulate and publish proposals for the preservation and enhancement of conservation areas, after consultation with local people at a public meeting. The advice insists that simply designating a conservation area will not ensure its protection and that

Legislation

planning authorities should analyse what makes an area special and develop policies to protect it, in consultation with local residents and businesses.

Other powers

The guidance gives powers to councils to control some minor developments in conservation areas including alterations to roofs, doors and windows, so as to prevent damage to the area's special character and to avoid councils having to pursue individual Article 4 directions. From October 1994 many churches, previously exempt, were brought within conservation controls although christian churches in active use still have ecclesiastical exemption.

Following publicity for the English Heritage list of battlefields, PPG15 reminds developers and councils to also protect wider features of the historic environment such as important gardens and parks, which are now separately listed and graded. World Heritage Sites have a high measure of protection.

Preservation presumption

PPG 15 re-affirms the presumption in favour of preserving listed buildings of special architectural or historic interest and a presumption in favour of retaining unlisted buildings which make a positive contribution to the character or appearance of a conservation area.

Environmental impact assessments

The principal environmental impact assessment (EIA) regulations are the Town and Country Planning (Environmental Impact Assessment) (England and Wales) Regulations 1999, as amended in 2000. An applicant may make a request for a pre-application screening opinion from the Council. Where a particular proposal is likely to have a significant environmental effect then a local planning authority may request an EIA for developments of the type listed in Schedule 2. Schedule 1 developments always give rise to EIAs. An applicant may apply for a screening direction (through a speedy process) to the ODPM if the applicant believes the EIA is unnecessary. A 16-week decision period applies to any EIA application, running from the deposit of the EIA with the local authority. From June 1995 any development requiring EIA loses the benefit of permitted development. See DTLR Circular 2/99.

Legal and lease

Legislation

➡ Dilapidations — 78

➡ Section 18(1) of the Landlord and Tenant Act 1927 — 83

➡ Latent Damage Act 1986 — 85

➡ Discovery of building defects – statutory time limits — 85

➡ Expert witness — 86

➡ Dispute resolution — 88

➡ The Provisions of Part II of the Housing Grants, Construction and Regeneration Act 1996 — 89

➡ Adjudication under the Scheme for Construction Contracts – how to get started — 92

Legislation

Dilapidations

The principle

Dilapidations is part of a legal procedure and the fundamental purpose of a Schedule of Dilapidations is to identify any breaches of covenant within a lease. The allegation of a breach of contract is the first step in the legal process. Consequently, for a Schedule of Dilapidations to have worth and perform the function for which it is drafted, it is necessary that the schedule is enforceable in a court of law.

The legal remedy for breach of contract is often a claim for damages and therefore the Schedule of Dilapidations is usually prepared as a claim for the cost of works.

Reform

New Civil Procedure Rules came into force in April 1999. A protocol has been drawn up by the Property Litigation Association (the most recent draft is dated July 2002) and it is hoped it will be submitted to the Lord Chancellor's Department to be approved under the Civil Procedure Rules (CPR). The underlying intention of CPR is to increase the number of pre-action settlements, reduce court time, expense and ultimately the extent of litigation. The Civil Procedure Rules have made a number of changes.

- ❖ Pre-action offer - this can be made by either party pursuant to Part 36 of The Civil Procedure Rules.
- ❖ All claims must be accompanied by a statement of truth signed by the claimant.
- ❖ Expert witness - the role has fundamentally changed under CPR. The expert's function is to give an independent expert view; he owes a duty to one body only, and that is the Tribunal before whom he appears. He is not there to create an argument in support of the party that appointed him.

The protocol

Although the protocol has not yet been approved under CPR, the courts may treat it as the normal and reasonable approach to pre-action conduct, and non-compliance might bring sanctions against the party concerned.

The protocol suggests a time limit of two months, after the end of the term for the landlord to have served the Schedule of Dilapidations. The tenant then has two months to respond and thereafter the respective surveyors should meet on site to discuss the claim, and seek to agree as many of the items in dispute as possible.

An early assessment of the claim is essential.

The claim should be drawn up as a separate document and should identify how it has been compiled, give a summary of the facts that the claim is based on, the VAT status of the landlord, all supporting documents and the date by which the tenant should respond.

The protocol clause 4.1.2 states that:-

- ❖ If the landlord has carried out the work then no Section 18 (1) Valuation is required.
- ❖ If the landlord intends to carry out the work, then it must state when the works are due, what steps have been taken towards doing them, and in most cases, it must provide a Section 18 (1) Valuation.
- ❖ If the landlord does not intend to carry out the works then it should provide a Section 18 (1) Valuation.

Legislation

Stages in the dilapidations process
Stage 1 – preparation

Obtain and appraise all relevant documentation including:
- leases;
- licences to alter;
- schedule of condition;
- side letters;
- photographs;
- fit out specifications;
- agent's letting brochures;
- any statutory notices served;
- deeds of variation;
- schedules of landlord and tenant's fixtures and fittings;
- details of outstanding service charges; and
- any rent deposit agreements.

This list is by no means exhaustive.

Stage 2 – the inspection

This must be comprehensive and thorough and include specialist professions if deemed necessary, such as a structural engineer or mechanical or electrical consultant.

Establish the original condition at the beginning of the term, and standard of repair that the tenant is required to undertake and identify the remedial work. Take into account the age, character and locality of the premises when let (Proudfoot Vs Hart 1890).

Include all measurements to aid calculation of the remedial works and use as proof as required at a later date.

Stage 3 – preparation of the Schedule of Dilapidations and claim

The schedule should contain the information shown in the first five columns below. The additional columns relating to the tenant's and landlord's comments turn the Schedule of Dilapidations into a Scott Schedule used as a negotiating tool between respective surveyors.

ITEM	CLAUSE No.	BREACH	REMEDIAL WORK REQUIRED	LANDLORD COST (£)	TENANT'S COMMENTS ON		LANDLORD'S COMMENTS ON	
					BREACH & REMEDY	COST (£)	BREACH & REMEDY	COST (£)

The Schedule of Dilapidations should be accompanied by a claim letter, which must include:
- landlord's and tenant's name and address;
- a clear summary of the facts on which the claim is based;
- the Schedule of Dilapidations (a separate document);
- any documents such as invoices and evidence of costs;

Legislation

- ❖ confirmation that the landlord and advisors will attend meetings;
- ❖ a date by which the tenant should respond; and
- ❖ a summary of the claim including:
  - cost of works
  - preliminaries
  - overheads and loss of profit
  - surveyors fees for preparing the Schedule (quantified and substantiated)
  - loss of rent
  - loss of service charge
  - surveyors fees for negotiating a settlement (projected)
  - if a Section 18 (1) cap applies then this should be shown.

Stage 4 – the response and negotiations

The Schedule of Dilapidations should be served within a reasonable time before the termination of the tenancy but not more than two months afterwards.

If a notice from the landlord has to be given to the tenant for reinstatement items then this must be served within a reasonable period before the end of the tenancy so that the tenant is still able to carry out these works in time.

Electronic copies of the Schedule of Dilapidations should be provided to facilitate easier negotiation, preferably in a Scott Schedule format as shown in the diagram on the previous page.

Following submission of the claim, the tenant must respond within a reasonable period, usually two months.

Surveyors should meet within one month of service of the Schedule of Dilapidations on a without prejudice basis preferably on site to establish the facts. If further meetings are necessary a strict timetable should be adopted.

Experts of respective parties in their like disciplines should also meet within one month of the tenant's response.

Proceedings should not be issued less than one month after the meeting of experts, or three months after serving of the schedule (whichever is the earlier).

Valuation – diminution in value

See Section 18(1) of the Landlord and Tenant Act 1927 on page 83.

Dilapidations v The Disability Discrimination Act (DDA) 1995

From October 2004, a provider of goods and services to the public will be obliged to take steps to change physical features within a premises where such features make other use of any service which is offered to the public impossible or unreasonably difficult.

The responsibility and duties in the DDA can rest on either a landlord or a tenant.

If a tenant, either in a single let building, or a multi-let building, provides a service to the public, the duties fall on the tenant.

If a landlord provides a service to the public, then they are responsible for any common parts of the building, whether it is a multi-occupied office building, or a shopping centre. However, it is not clear who is responsible for the common parts in the case where the public only visit the building at the invitation of the tenant.

Where a tenant has undertaken works to comply, there are implications

Legislation

under the Disability Discrimination (Providers of Services) (Adjustments of Premises) Regulations 2001. If a landlord wishes to remove the alteration, there is nothing to prevent him doing this after the outgoing tenant has vacated, but all reinstatement of these elements must be at his own cost. Any requirement on the outgoing tenant to remove these alterations may be considered unreasonable and could be challenged in the courts. This is as yet untested, and it remains to be seen whether a tenant will be able to use this in defence against a reinstatement claim.

Service charge

One of the most important aspects of maintaining a landlord's investment in a property, is to ensure that it is kept in repair. In a multi-occupied building, let on full repairing terms, the landlord will seek to recover costs of all repair and maintenance work from various tenants through a service charge.

An important case, Fluor Daniel Properties Limited v Shortland Investments Limited (2001), encompasses some already established principles relating to service charge recovery.

In order for a landlord to carry out repairs to a multi-occupied property, and recoup the cost from the tenants, he must establish that the item in question is out of repair, and to do this, a number of circumstances must be taken into account. These include the nature and life span of the building, the length of lease, and the extent of the defect. The courts also state that a landlord must have regard to the benefit that a tenant would derive from the repairs.

The question of standard relies on a number of circumstances as laid down by Proudfoot v Hart (1890). The standard must have regard to the age, character and locality of the building and whether the condition of the subject matter is reasonably acceptable to a reasonably minded tenant, of the kind likely to take on a lease of the building or part thereof.

The landlord must have proper regard to the interests of the tenants, and in some circumstances may carry out improvements. As the environment in which the building operates is improved, a reasonable tenant may expect features to be improved such as high speed lifts, or better performing air conditioning.

The landlord must act reasonably and the obligation to repair must not be looked at on its own; in every instance it is a question of degree.

Break clauses

These are covenants within a lease that give the option for the landlord or the tenant to bring the lease to an end before the contractual expiry date. It is important to establish whether they are condition precedent or not.

Where they are conditional, the tenant may be required to comply with certain conditions before the break option can be successfully exercised. This may include full compliance with all repairing obligations within the lease. Failure to comply may result in the lease continuing for the remainder of the term.

VAT and dilapidations

Protocol, Clause 4.2, requires that the VAT status of the landlord should be stated. VAT in respect of dilapidations is a complex subject with its own set of rules, regulations and case law. Opinions are usually divided and much will depend on the circumstances.

Where a tenant is making a payment to a landlord in full and final settlement of its dilapidation liabilities, under Customs and Excise rules, the payment is not a 'taxable supply' for the purposes of VAT. This is because Customs and Excise deem the payment to be one of damages and not a supply of something. VAT may be payable if the landlord passes the payment on to an incoming VAT registered tenant.

Legislation

Complying with European Community Law, the Finance Act 1989 introduced major changes that included giving UK payers of VAT the option to pay VAT on supplies relating to an interest in commercial land. Where a person or company is registered for VAT there is a special statutory exemption to the charging of VAT on such supplies. The tax payer can waive this exemption if they desire. To do this they must notify HM Customs and Excise. For clarity, if a landlord has elected to waive exemption on a building, he must charge VAT on such items as rent received (ie supplies that the tenant is receiving). In this instance, it will not be appropriate to include VAT as part of a dilapidations claim. If the VAT exemption has not been waived then VAT need not be charged by the landlord on supplies, and VAT may form part of a dilapidations claim.

So the first issue to establish is whether the landlord is registered for VAT. Is the landlord required by VAT law to charge VAT on its normal business transactions, such as rent receipts. If so, has it elected to waive VAT exemption on the specific building that is subject to the dilapidations claim?

Once this has been established, a simplified VAT analysis might follow the steps shown in the diagram.

The protocol requires that the claim establishes what the landlord's intentions are, and states the landlord's and/or the demised property's VAT status. A basic understanding of the VAT issues is therefore necessary, if the requirements under the protocol for an accurate initial claim are prepared.

Dilapidations VAT Analysis

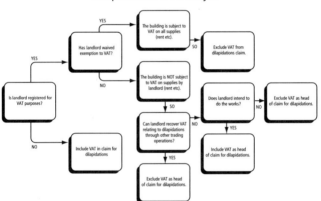

Summary of important statutes

Landlord and Tenant Act 1927 Section 18 (1)
Limits the cost of a claim for breach of covenant (more details on this act are shown opposite).

Law of Property Act 1925 Section 146
Prescribes the form of notice for re-entry for forfeiture.

Law of Property Act 1925 Section 147
Provides relief for tenants on long leases in respect of internal redecorations.

Legislation

Leasehold Property (Repairs) Act 1938
Gives protection to certain tenants in respect of Section 146 Notice (ii) above.

Defective Premises Act 1972
Provides that a landlord shall be liable for lack of repair in cases where he/she knew or ought to have known of the defect.

The Civil Procedure Rules 1998
Introduced by Lord Woolf, they provide rules and practice directions for dispute procedures.

Summary of important case law

There have been a number of leading decisions relating to dilapidation law in the last few years. The following cases give an indication of developing areas of law.

Scottish and Mutual Assurance Society Limited v British Telecommunications plc (1999) (E.G.C.S 43)
Section 18 (1) of the Landlord and Tenant Act 1927 Part II.

Loss of rent.

Notice for reinstatement of alterations.

Shortlands Investments Limited v Cargill plc (1995) (E.G.L.R 51)
Section 18 (1) of the Landlord and Tenant Act 1927 Part II.

Trane (UK) Limited v Provident Mutual Life Assurance Co Limited (1995) (W.G.L.R78)
Compliance with conditions of break clauses.

Jervis v Harris (1996) (E.G.L.R 78)
Use of provision for landlord's re-entry.

Extent of recovery of expenditure.

Mannai Investments Co. Limited v Eagle Star Life Assurance Co Limited (1997) (E.G.L.R 69)
Accuracy of notice showing intent to break tenancy.

Credit Suisse v Beegas Nominees Limited (1994) (4All,ER803)
Establishing the interpretation of the repairing covenants and the different types of obligation placed on the tenant.

Section 18(1) of the Landlord and Tenant Act 1927

Section 18(1) of the Landlord and Tenant Act 1927 is divided into two Limbs.

> **Limb 1,** *provides that damages for a breach of covenant ..... to put or keep premises in repair ..... shall in no case exceed the amount ...... by which the value ..... of the premises is diminished owing to the breach.*

Limb 1 of Section 18(1) applies in all dilapidations cases.

Legislation

The common law claim in dilapidations, usually comprises the following components:-

1. Cost of works.
2. Fees for preparation and service of the Schedule.
3. Fees for carrying out of the works.
4. Fees for negotiating the claim.
5. Loss of rent, loss of empty rates.
6. Loss of insurance.
7. VAT.

Limb 1 of the Section 18(1) provides that whatever the amount of the common law claim, the landlord's entitlement in damages is limited to the amount by which the value of his interest has been diminished owing to the breach.

The operation of the Act requires comparison to be made between two valuations:

Valuation A - Assuming compliance.
Valuation B - Assuming actual condition.

The difference between Valuation A and Valuation B assesses the statutory cap on the amount of damages payable.

It has been held that where Valuation A is a negative figure and Valuation B is a greater negative figure the difference, amounting to the landlord's loss, is payable (Shortlands Investments Plc -v- Cargill Plc).

> Limb 2 of Section 18(1) further provides, ... no damages shall be recovered ..... if it is shown that the premises, in whatever state of repair they might be or that shortly after the termination of the tenancy have been or be pulled down or such structural alterations made therein as would render valueless the repairs .....

Limb 2 of Section 18(1) would operate, for example, where it could be demonstrated that the landlord of an unmodernised 1960s office building intended, as at the expiry of the lease, to carry out refurbishment works to install a suspended ceiling, a raised floor and air conditioning.

In such a case, the landlord's proposals would significantly impact upon the repairs required to the interior of the building, rendering the benefit of the required repair valueless. As a consequence, to the extent that this work is rendered valueless, it will fall out of the claim.

However, unless the landlord intends to carry out works to the exterior of the building that will render valueless the exterior disrepair, the repairs required will remain a valid part of the claim.

The critical date upon which to establish the landlord's intention is at the expiry of the lease (see Salisbury -v- Gilmore (1948) and Cunliffe -v- Goodman (1950)

The fact that the landlord is contemplating a number of options, including a potential refurbishment, as at the expiry of the lease is not a sufficient basis itself to render the value of the repairs nugatory. A clear and fixed intention must be demonstrated.

Section 18(1) relates to repair only. It does not relate to the reinstatement or decoration content of the claim. However, these elements of the claim are governed by common law principles under which the landlord's entitlement is to be reimbursed for his loss, and the same principles apply. It will be relevant however, in assessing the loss, whether the landlord truly intends to reinstate tenant's alterations, for example, a mezzanine floor in a warehouse.

The introduction of the Pre Action Protocol for Dilapidations has confirmed that valuations under Section 18(1) are required when a claim is made in virtually all cases. The exceptions to this rule are when the landlord has either carried out the works or can demonstrate intention to do the works.

As a consequence, both landlords and tenants should obtain advice as to the likely impact of Section 18(1) on their claim at an early stage.

Latent Damage Act 1986

The Latent Damage Act 1986 enhances case law relating to certain cases of negligence, by imposing statutory limits in relation to the time in which cases may be brought to court if a defect is found.

The Act applies specifically to those cases of negligence that relate to latent damage and not to personal injury. In addition, recent case law suggests that in the meaning of the Act, negligence covers only a breach of a tortious duty of care and not a contractual duty of care.

Previous statute exists indicating time limits in which legal action must commence; section 14 of the Limitation Act 1980. The 1986 Act enhances section 14 by limiting any claims to six years after the damage occurred. This, however, can be extended under section 14A by a further three years from the date when the defect is discovered. In all cases however, a 15-year time limit is placed under section 14B of the Act. When considering claims for latent defects, the most important points to establish are the dates from when the negligence occurred and the cause of the negligence. For example, the negligence may have been caused by defective construction, design, or a combination of both. The date on which the defect technically arises may be confused by the ongoing contractual state of the project, ie whether a final certificate has been issued or a defect liability period is still active. In certain cases proceedings may be delayed on the basis that the damage has not yet occurred, but is in fact imminent.

Discovery of building defects – statutory time limits

The Limitation Act 1980 ('the 1980 Act') as modified by the Latent Damage Act 1986 (see above) lays down the periods within which proceedings to enforce a right must be brought. Upon the discovery of a defect in a building or structure, possible claims against the designers and constructors of that building or structure could arise in contract, in tort or under statute.

Contract

To bring an action for breach of a simple contract, court proceedings must be commenced within six years of the date on which the breach of contract occurred. If the contract has been completed under seal, rather than under hand, then the time limit prescribed by the 1980 Act is 12 years. Under the Companies Act 1985 (as amended) the 12-year limitation period also applies to companies when a contract is signed as a deed by two directors or a director and a company secretary.

Tort

The general time limit for actions in tort is six years from the date when the damage was suffered, with the exception that a time limit of three years applies to personal injury actions involving negligence. This period runs from the date of the accident concerned, although special rules apply for illnesses, which may not manifest themselves for many years after exposure to their cause, for example, asbestos.

Also, for negligence claims involving latent damage, the time limit laid

Legislation

down by the Latent Damage Act 1986 for commencing proceedings is six years from the date the damage was suffered. However this period can be extended for a further three years from the 'starting date'. The starting date is defined by the Latent Damage Act 1986 as the date on which the plaintiff first had both the required knowledge and the right to bring an action. The Latent Damage Act also includes a 'long-stop' provision preventing the instigation of any proceedings after the expiry of 15 years from the date of the negligence concerned.

Statute

Where a right derives from a breach of statutory duty, reference should, in the first instance, be made to the particular statute concerned, which may specify a time limit for the commencement of proceedings. The Defective Premises Act 1972 provides a particularly germane example. If the statute concerned is silent as to the time limit, then a period of six years will generally apply. If an action is brought in respect of a defective product, under the Consumer Protection Act 1987 there is a cut-off point of ten years from the 'relevant time' (usually the date of a supply).

It should be noted that the time limits for both tortious and contractual claims might be postponed where there is concealment, mistake or fraud.

Expert witness

Part 35 of the Civil Procedure Rules (CPR) which came into force on 26 April 1999 deals with expert witnesses.

The CPR states, "it is the duty of an expert to help the courts on matters within his/her expertise. This duty overrides any obligation to the person from whom he/she has received instructions or by whom he/she is paid". The expert should therefore:

- be independent/impartial;
- state any reservations about the case he/she is instructed upon;
- identify areas outside his/her expertise;
- consider all material facts in his/her report and state all facts and assumptions, upon which his/her opinion is based;
- state where necessary that it is only a provisional report, because only limited information/data was available when it was compiled;
- advise instructing solicitors if further information, ie the other expert's report, changes his/her opinion;
- make available all documents referred to in the report, ie survey reports, plans, calculations, photographs etc; and
- keep the report as brief as possible, but without losing the reasoning and conclusions, upon which his/her opinion is based.

The written report

There are a number of model forms of report, for example, the model form produced by the Academy of Experts. The CPR lays down formal requirements in the practice direction that accompanies it.

Under the new rules, there needs to be a statement setting out the brief and instructions given.

The report is to be written in the first person and it is an individual who prepares the report and not the company or firm.

The report should be addressed to the court.

Legislation

Although the report should be as brief as possible, accuracy should not be sacrificed to brevity.

The expert witness must be able to substantiate each and every sentence of the report and highlight any areas where his/her opinions are based on inadequate factual information. It is not the expert's role to make or advance legal arguments.

The report should contain his/her curriculum vitae (CV).

It is compulsory under the new rules to include a declaration in the report that the expert understands that his/her duty is to the court and that he/she has complied with that duty. A statement of truth must also verify the report. The wording to be included within the expert's report immediately before the signature is as follows:

> **I believe the facts as stated in this report are true and that the opinions I have expressed are correct.**

Under the CPR, each party has 28 days after receipt of the opposing expert's report to put written questions. Unless the court gives permission for more general questions, these can only be for the purpose of clarifying the report.

Without prejudice meetings

Without prejudice meetings between experts are necessary and important. The court, under the CPR, may, and normally does, require the experts to produce a joint statement from 'without prejudice' meetings setting out what has and has not been agreed.

Giving evidence in court

If the case gets to a hearing, then the expert witness will be required to give evidence. The stages of examination of the evidence will be:

- examination in chief;
- cross-examination;
- re-examination; and
- questions from the judge.

A few helpful hints in giving evidence are listed below:-

- Take time, don't rush.
- Succinctly answer only the questions that are asked.
- Use plain language.
- Do not digress from the question asked.
- Do not act as advocate.
- If the question is not understood nor heard – say so.
- Know the report 'inside out'.

Cross-examination will challenge credibility; so consider the following:-

- Do not feel obliged to fill a silence.
- Do not be afraid to answer the same question again and again.
- If asked a closed question, then state that there may not be a yes or no answer.
- Do not be rattled by a number of quick fire questions.
- Do not argue with counsel.
- If he/she becomes aggressive, stay cool.
- Beware of the important question slipped in among a number of trivial questions.

The above notes only deal briefly with the Civil Procedure Rules. These notes should not be considered as comprehensive text. The role of the expert is evolving through the interpretation of the CPR in case law and

Legislation

any expert should ensure that he/she fully understands that role in the light of the current law.

Dispute resolution

Disputants and their advisors now have a wide variety of dispute resolution mechanisms that they can select to resolve their disputes.

The principal dispute resolution procedures are:

Adjudication
This process is now enshrined in the Housing Grants Construction and Regeneration Act 1996 (Act). A wide variety of the disputes arising under construction contracts can be referred to adjudication. The Act does not deal with residential disputes, but the JCT Consumer Contracts contain adjudication clauses. Adjudication is a popular method of dispute resolution not least because an adjudication award has to be made within 28 days of the case being referred to an adjudicator.

Arbitration
Arbitration is based on the contractual relationships between the parties themselves and between them and the arbitrator. New procedures under the 1996 Arbitration Act has given the arbitrator wide powers to resolve disputes without unnecessary cost or delay, and in a fair manner without undue interference from the courts.

Early Neutral Evaluation (ENN)
Technology and Construction Court judges are prepared to arrange a short hearing of a case, or specific issues in it, on a without prejudice basis and give preliminary views on its merits, as an aid to settlement discussions between the parties. If a judge determines a particular issue by ENN the parties are free to agree whether or not they will be bound by it. If the ENN does not result in settlement, the case can proceed to trial but will be heard by another judge with no knowledge of the outcome of the ENN.

Independent expert
This forms a very valuable means for the speedy resolution of technical disputes. The procedure is generally straightforward and flexible. Issues in dispute are referred to an expert to decide using his/her or her own professional expertise or judgement. It has been successfully used for many years in rent review matters, but has much wider application to technical disputes. If the parties agree to be bound by the expert's decision it cannot be appealed unless there is misconduct on the part of the expert.

Litigation
The CPR has led to more efficient running of cases both in terms of costs and time as the court now becomes directly involved in case management. The Technology and Construction Court in the High Court has considerable experience of dealing with construction disputes. Before court proceedings are commenced, the parties should comply with the pre-action protocol for construction and engineering disputes. This requires the parties to set out their respective cases in correspondence and to meet on a without prejudice basis to seek to settle the dispute or narrow the issues involved, prior to the issue of court proceedings.

Mediation
Mediation is a voluntary and non-binding procedure. It is a private process in which an independent neutral person helps the parties reach a negotiated settlement.

Legislation

The Provisions of Part II of the Housing Grants, Construction and Regeneration Act 1996

The following aims to set out a brief outline of the issues which need to be considered when determining whether contractual terms are compliant with Part II of the Housing Grants, Construction and Regulation Act 1996 or whether certain provisions will be incorporated into the contract by the Scheme for Construction Contracts (Scheme).

Parties to a construction contract are free to negotiate and agree the terms and conditions under which the works and services are to be carried out. However, there are times where a contract fails to comply with the minimum requirements relating to adjudication and payment laid down by the Act. Consequently, certain provisions will automatically be incorporated into the contract by the Scheme for Construction Contracts.

When does the Act apply?
(S104 - 107)

The Act applies to:-

- ❖ Contracts entered into on or after **1 May 1998.**
- ❖ **Contracts in writing.** It is sufficient if the contract is evidenced in writing.
- ❖ Contracts for **construction operations** which include the construction, alteration, repair, maintenance, decoration, demolition and installation in buildings forming or to form act of the land and also architectural, design, surveying or engineering advice.
- ❖ The carrying out of construction operations in England, Wales or Scotland whatever the applicable law of the contract.

The Act does not apply to:-

- ❖ Contracts with **residential occupiers** for work on their property where they intend to occupy the property as their residence.
- ❖ Certain mining, drilling and extraction operations.
- ❖ Installation or demolition of plant or machinery or steelwork to support or provide access to plant or machinery on a site where the primary activity is nuclear processing, power generation, water or effluent treatment, or the production processing of chemicals pharmaceuticals, oil, gas, steel, food or drink.
- ❖ Manufacture and delivery of materials not involving installation.
- ❖ Artistic works.

Letters of intent may be subject to the Act where they are sufficient to amount to a legally binding contract in their own right.

Adjudication (S108)

The contract must:-

- ❖ Allow either party to refer a dispute to adjudication at any time.
- ❖ Provide for the appointment of an adjudicator within seven days.

Legislation

- ❖ Require the adjudicator to reach his decision within 28 days **after** the dispute has been referred to the adjudicator (or a longer period if agreed by the parties).
- ❖ Allow the adjudicator and the party who referred the dispute to him to extend the period for his decision by up to 14 days.
- ❖ Impose a duty on the adjudicator to act impartially.
- ❖ Enable the adjudicator to take the initiative in ascertaining the facts and the law surrounding the dispute.
- ❖ Provide for the decision of the adjudicator to be binding on the parties until the dispute is taken to arbitration or the Courts.
- ❖ Provide that the adjudicator is not liable for anything he does unless he acts in bad faith.

If the contract does not comply with all eight elements in full, all the adjudication provisions of the contract will be set aside and the adjudication procedures under the Scheme for Construction Contracts will apply. The procedures under the Scheme cover the eight points listed above and introduce time limits.

Payment

The Act and the Scheme for Construction Contracts provide a 'menu' of payment provisions covering:

- ❖ payment by instalments;
- ❖ final payment;
- ❖ withholding payment; and
- ❖ conditional payment.

If a contract fails to comply with any one of the provisions from the 'menu' the relevant provisions from the Scheme will apply. The remainder of the contractual provisions that do comply with the Act will remain intact. It is therefore possible to end up with a contract where the payment provisions are a mixture of express terms agreed between the parties and implied terms from the Scheme.

Payment by instalments (S109 - 110)

A party to a construction contract is entitled to payment by instalments, stage payments or other periodic payments unless:-

- ❖ The contract specifies that the duration of the work is less than 45 days; or
- ❖ The parties agree that the work is estimated to take less than 45 days.

Where the work falls within the 45 day limit, the right to instalment payments is excluded but all other payment provisions (notice of withholding payment, set-off, etc) will apply as will the adjudication provisions outlined above.

Where a contract falls below the 45 day limit, payment of the contract price falls due 30 days after completion of the work (or 30 days after the contractor's claim if later) and payment must be made within 17 days.

In all other cases, the parties can agree between themselves:

- ❖ the amounts of each payment;
- ❖ the intervals between each payment;
- ❖ the date each payment becomes due; and
- ❖ the final date by which each payment must be made.

Legislation

If the contract does not contain a clear mechanism for determining each of these four elements they will be determined by the Scheme for Construction Contracts, namely:-

- ❖ The amount of each payment will be based on the value of the work and other costs to which the contractor is entitled during the payment interval.
- ❖ There will be 28 day payment cycles.
- ❖ Payment is due seven days after each 28 day period (or seven days after the contractor's claim for payment if later).
- ❖ The final date for each payment is 24 days after each 28 day period (or 24 days after the contractor's claim for payment if later).

The contract must provide for the paying party to give notice within five days of the date on which each instalment becomes due, specifying the amount proposed to be paid and the basis on which it is calculated. Any attempt in the contract to vary or exclude this requirement will be ineffective and this provision will be implied by the Scheme for Construction Contracts.

Final payment

The contract must contain a clear mechanism for determining when the final payment due under the contract becomes payable and the final date by which that payment must be made.

The parties are free to agree the dates or periods within which the final payment is due and is payable but if there is no such mechanism, in accordance with the Scheme the final payment:

- ❖ is due 30 days after completion of the work (or 30 days after the contractor's claim for payment if later); and
- ❖ the final date for making the final payment is 47 days after completion of the work (or 47 days after the contractor's claim for payment if later).

Withholding payment (S111)

- ❖ No payment can be withheld unless a 'notice of intention to withhold payment' has been given, specifying the amount to be withheld and the grounds for withholding payment.
- ❖ The notice must be given before the final date for payment.
- ❖ The contract can specify how long before the final date of payment, the notice must be given (even if it is just one day) but if this notice period is not specified, the Scheme applies and at least seven days notice must be given.

Conditional payment (S113)

Any provision in a contract which makes payment conditional upon the paying party receiving payment from someone else is ineffective and the payment provisions of the Scheme outlined above will apply.

The only exception is where the contract provides that payment may be withheld if the reason for non-payment is the insolvency of someone else in the payment chain.

Suspending performance (S112)

If any payment is not received by the final date for payment and a notice of withholding payment has not been served, the contractor may suspend work after giving seven days notice of his intention. The right to suspend performance ceases when the relevant payment is received. The period in

Legislation

which to complete the works is automatically extended by the number of days of the suspension.

This brief summary of the provisions of Part II of the Housing Grants, Construction and Regeneration Act 1996 is not intended to be a detailed explanation of the provisions of the Act and we recommend that legal advice is sought on any specific issues.

Adjudication under the Scheme for Construction Contracts – how to get started

Adjudication under the Housing Grants, Construction and Regeneration Act 1996 (Act) allows for a quick fix method of dispute resolution. The right to refer a dispute to adjudication is available to a party to a construction contract within the meaning of the Act at any time.

Below is a brief explanation of the steps required to commence an adjudication under the Scheme for Construction Contracts, ie **where there are no contractual adjudication provisions.**

Where adjudication is your chosen method of dispute resolution, it is essential that you comply with the strict time limits laid down by the scheme and any timetable imposed by the adjudicator. The adjudicator is under an obligation to reach a decision within 28 days of referral unless the parties agree otherwise.

It is not possible to contract out of the time constraints laid down for adjudication by the Act.

Procedure

To commence adjudication you must take three steps:-

- ❖ Give notice of adjudication.
- ❖ Request an adjudicator to act.
- ❖ Serve a Referral Notice.

Notice of Adjudication

This must be in writing, be given to every other party to the contract and contain:

- ❖ details of the parties involved;
- ❖ a brief description of the dispute;
- ❖ details of when and where the dispute arose;
- ❖ what you are seeking from the adjudicator, for example, an award for a specific sum; and
- ❖ names and addresses of the parties to the contract (including the addresses which the parties have specified for the giving of notices, if any).

Note: In view of the very tight timescale for adjudication you should ensure that your claim is fully prepared before issuing the Notice of Adjudication.

Appointing an adjudicator

After giving notice of adjudication you must make a request for an adjudicator to act. The timescale for the appointment of an adjudicator is extremely tight and a request for the appointment of an adjudicator should be made at the same time as giving the Notice of Adjudication. In

Legislation

order to determine who to appoint you should consider the following:-

If your contract names an adjudicator, contact him/her to ensure that he/she is ready and willing to act.

If an Adjudicator Nominating Body (ANB) is named in your contract, contact that body and ask for an appointment to be made.

If no adjudicator or ANB has been named in your contract you can contact any ANB such as:-

The Academy of Independent Construction Adjudicators, 020 7608 5221.
The Royal Institution of Chartered Surveyors, 020 7222 7000.
The Chartered Institute of Arbitrators, 020 7837 4483.
The Royal Institute of British Architects, 020 7580 5533.
The Technology and Construction Solicitors Association, 020 7655 1000.
The Construction Industry Council, 020 7637 8692.

Contact the most appropriate ANB depending upon the nature of the dispute and the issues involved. Some ANBs will be able to offer a greater diversity and breadth of experience compared to other single discipline organisations.

If an ANB is used: The ANB has five days to inform you of the nominated Adjudicator. The nominated adjudicator then has up to two days to confirm his appointment.

Caution: If a named or nominated adjudicator refuses to act, another adjudicator can be agreed or nominated by any ANB, but beware the time limit for issuing the Referral Notice. If an alternative adjudicator cannot be appointed within seven days of the Notice of Adjudication the safest course is to issue a fresh Notice of Adjudication.

Referral Notice

Within seven days of the Notice of Adjudication you must send the Referral Notice to the adjudicator formally referring the dispute to him. The Referral Notice must be in writing, be given to the adjudicator and every other party to the dispute and:

- ❖ contain the basis of your claim, including an explanation of how the dispute arose and identifying the issues in dispute;
- ❖ be accompanied by copies of (or relevant extracts from) your contract (whether this is a standard printed form or evidenced in correspondence);
- ❖ include any documents upon which you wish to rely in support of your case;
- ❖ contain the remedies and award you are seeking; and
- ❖ give the adjudicator wide jurisdiction by giving him/her an alternative, for example, "such other sum as the adjudicator may determine".

It is important that the Referral Notice clearly sets out the issues in dispute which the adjudicator is being asked to determine including the history of the case and any arguments raised by the other party as well as identifying the remedies sought. Claims not identified in the notice of Adjudication cannot be introduced later in the same adjudication.

Following service of the Referral Notice the adjudicator should set a timetable for dealing with the adjudication including a response from the opposing party and request any further information or evidence in order to reach his decision.

Neighbourly matters

Legislation

→	Rights to light	96
→	Daylight and sunlight amenity	101
→	Party wall procedure	102
→	Access agreements	105
→	Construction noise and vibration	106

Legislation

Rights to light

The subject of rights to light usually concerns the assessment of whether proposed obstructions (for example, new developments) are likely to interfere materially with neighbours' easements of light. Interpretation of whether a material right to light issue is likely to arise requires knowledge of the law, particularly easements and nuisance, and an understanding of the technical measurements of skylight entering a room. It is then possible to assess the risk of injunction and, where appropriate, the likely level of damages that could be awarded. It is also possible to determine how to modify a proposed scheme in order to reduce or overcome potential problems.

Legal background

Rights to light problems bring together two distinct but different areas of English law namely private nuisance - a sub-division of the law of torts; and easements - a sub-division of land law.

Private nuisance

The tort of private nuisance, like the tort of public nuisance, regulates activities affecting individual rights in or rights over real property (land).

A private nuisance may be defined as:

> An unreasonable interference with a person's use or enjoyment of land itself, or some right over or in connection with land (ie a right to light).

The law of nuisance tries to balance the legitimate activities of neighbours - a give and take approach. Interference with a right to light must be objectively unreasonable in its extent and severity if it is to be sufficient to constitute a nuisance in the eyes of the courts. Only when the courts are satisfied that the interference is unreasonable will they remedy the situation by awarding an injunction and/or damages. It should be appreciated that levels of natural light can often be interfered with to a marginal extent and this will not necessarily constitute an infringement of a proprietary right that will be recognised as a nuisance.

Easements

An easement may be defined as a right annexed to land to use or to restrict use of neighbouring land in some way. More particularly, for a right to exist as an easement (as opposed to a restrictive covenant) the right must possess the four essentials:-

- There must be a dominant tenement and a servient tenement.
- The right must accommodate (benefit) the dominant tenement.
- The dominant tenement and the servient tenement must be owned or occupied by different persons.
- The right concerned must be capable of forming the subject matter of a grant.

Rights to light satisfy the aforementioned essentials and, like some other classes of right, they have existed as easements for many centuries. The general rights to light principles set out below are distilled from the large body of case law that exists.

A right to light can be defined generally as:

> A negative easement providing a right for a building to receive sufficient natural light through a defined aperture (usually a window) in perpetuity or for a term of years.

Legislation

Nature of a right to light

A right to light is not personal - it runs with property/buildings.

A right benefits the dominant tenement and burdens the servient tenement, but such a right will not necessarily be permanent. A right to light is for 'sufficient' natural light only and this is taken to mean enough light, "according to the ordinary notions of mankind" for:

- ❖ comfortable use and enjoyment of a dwelling house; or
- ❖ beneficial use of and occupation of a warehouse, shop or other place (office, etc).

The test for 'sufficiency' is whether or not the dominant tenement will be left with enough light according to the ordinary requirements of mankind. Sufficiency is not based on the measure of light lost. See Colls v Home & Colonial Stores [1904] AC 179.

Actionable injury and measurement of light

No specific rule has been developed by the courts to define exactly when a reduction in natural light becomes actionable. The test for injury is uncertain but flexible. The court will have regard, in all cases, to the specific facts and circumstances and it will usually hear objective technical evidence from a rights to light expert, as well as more subjective evidence from the injured party.

A form of technical evidence has evolved that entails analysis of the amount of the notional sky dome that can be seen from a series of points in an affected room at table level. At any given point on the working plane there is a minimum amount of sky area below which the level of daylight at that point will be inadequate. Adequacy is considered to be just enough for undertaking work that requires visual discrimination, such as reading, drawing or sewing. In technical terms this is one lumen or 1/500th of a standard uniform dome of overcast sky in December (ie 0.2% sky factor).

It is possible to plot the 0.2% sky factor contour in the subject room and measure the area of the room that will receive more than adequate light both before and after development. Today, leading rights to light experts tend to do this with the aid of 3D computer modelling and specialist software, which is more accurate and efficient than the laborious Waldram method.

Having measured the area that will be adequately illuminated, it is possible to assess whether an actionable injury will arise. Very generally, for day to day practical purposes, light specialists have adopted the general conventions that:

- ❖ A commercial property should be considered actionably damaged when less than 50% of an office floor area is lit to the critical one lumen (0.2% sky factor) level.
- ❖ A residential/domestic property should be considered actionably damaged when less than 55% of a room area is lit to the critical one lumen (0.2% sky factor) level.

It must be understood that the percentages mentioned above are not strongly founded in specific legal authority. The courts regard the percentages as good guides but not absolute tests.

Acquisition of a right to light

A right to light may be created by:

- ❖ express grant or reservation (sometimes encountered);
- ❖ implied grant or reservation (rarely encountered); or
- ❖ prescription (very common and often called 'ancient lights').

Prescription means the procuring of a right on the basis of a long established custom and three methods of prescription exist, namely:

Legislation

- ❖ time immemorial (right enjoyed since before 1189);
- ❖ doctrine of lost modern grant (right enjoyed continuously for minimum 20 years); and
- ❖ Prescription Act 1832; Sections 3 and 4 (right enjoyed continuously for 20 years).

(Note: In the City of London, because of the 'Custom of London', the acquisition of a right by the doctrine of lost modern grant is not available - see Bowring Services Ltd v Scottish Widows 1995).

Defeating a right to light

Prescription through the doctrines of lost modern grant or time immemorial can be defeated if it can be shown that the easement has not been enjoyed, 'as of right', ie through force, secrecy or with permission.

Statutory prescription, under the Prescription Act 1832, may be defeated if the servient party can show that:

- ❖ at some time within the last 19 years, they have prevented the entry of natural daylight through the subject apertures by erecting an opaque physical obstruction for a continuous period of at least one year;
- ❖ at some time in the last 19 years, they registered a light obstruction notice under the Rights of Light Act 1959 for a whole year (see below); or
- ❖ the right has been enjoyed under some consent or agreement expressly given for that purpose by deed or in writing.

Defending your right to light

A prescriptive right to light may be lost if it is not defended in the face of development. A neighbour who acquiesces in or submits to an interruption of light for one year or more will lose their claim to a prescriptive right under section 4 of the Prescription Act 1832.

See Dance v Triplow and Another [1992] 17 EG, 103. Successful defence of a prescriptive right to light relies much on eternal vigilance and prompt protestation in writing to the obstructor. The protestations should also be repeated at regular intervals. Ultimately legal proceedings will need to be brought against a developer who ignores the objections and the timing of this may considerably affect the chances of obtaining an injunction.

Remedies

The current legal system permits the awarding of:

- ❖ prohibitory or mandatory injunctions; and
- ❖ common law damages.

Injunctions

The courts are reluctant to permit a servient party to buy himself out of a breach of covenant and, generally, an injunction is regarded as the normal remedy, with damages the exception. The court may award damages in lieu of an injunction if all the four tests below can be answered in the affirmative:-

- ❖ Is the injury small?
- ❖ Would a small money payment be an adequate remedy?
- ❖ Would it be oppressive to the defendant to grant an injunction?
- ❖ Is the injury one that can be estimated in money terms?

These tests are derived from Shelfer v City of London Electric Lighting Co [1895] 1 Ch 287 31.

Legislation

Damages (compensation)

Compensation for injury to a right to light may be assessed using one of two methods:

- ❖ the traditional valuation approach; or
- ❖ the developer's profit approach.

For commercial property, the valuation approach is based on a freeholder in possession. A 'base book value' is calculated for the light loss, which is then enhanced by a multiplier of up to three or four times, having regard to the case of Carr Saunders v Dick McNeil Associates Ltd and Others [1986]. Any compensation will be apportioned between the various interests in the dominant tenement. For residential property, the assessment of compensation is more subjective.

For very serious rights to light injuries, it may be more appropriate to adopt the developer's profit approach, when assessing compensation. This approach considers the possibility of sharing the profit that a developer will make from the extra floor space that could be prevented from being built if the adjoining owner obtained an injunction. The profit share can be up to 50/50 in very severe cases.

Dos and don'ts

- ❖ Do establish whether surrounding properties enjoy rights to light, including other tenanted parts of the client's property, and identify all parties with an interest.
- ❖ Do establish whether development proposals are likely to leave the surrounding buildings with inadequate light.
- ❖ Do take specialist advice from a rights to light consultant at an early stage.
- ❖ Do obtain copies of all leases, deeds, transfers, restrictive covenants, and so on, that could have a bearing on the legal position.
- ❖ Don't be fooled by buildings that look less than 20 years old or have blocked-up windows: case law is complex and there could still be a right to light!
- ❖ Do consider whether the Crown has ever had an interest in the development site: the surrounding buildings may not be entitled to rights to light.
- ❖ Do consider whether the development site has ever been acquired or appropriated by the local authority for planning purposes under section 237 of the Town and Country Planning Act 1990, as this affects the potential for injunctions.
- ❖ Do obtain copies (if available) showing the massing and profile of the existing building on the development site, and the proposed building or extension, and also up-to-date floor layout plans for the surrounding buildings.
- ❖ Do establish the extent of a right (ie the number, size and location of apertures).
- ❖ Do consider whether transferred or 'incorporated' rights to light exist.
- ❖ Do seek the advice of lawyers if the legal position is complicated beyond your experience by any controlling deeds or similar documents.
- ❖ Do ensure that your client understands that the law relating to rights to light, and the valuation techniques, are not an exact science.
- ❖ Do remember that you cannot rely on a neighbour, particularly a residential owner, settling for compensation.

Legislation

- ❖ Do explain to your client that there is no statutory procedure for dealing with rights to light issues and that there are no prescribed periods and deadlines during or by which parties are obliged to settle issues.
- ❖ Don't allow a 'dominant' party to be pressured into early agreement of compensation.

Light obstruction notices

Light obstruction notices can be used to either defeat an existing right to light that has been acquired by statutory prescription or prevent such a right from being acquired. They are a useful tool for preserving the development potential of a site and can also be used to 'flush out' potential claims.

Under Section 2(1) of the Rights of Light Act 1959 a light obstruction notice can be registered with the local authority as a local land charge for a period of one year.

Further information

The two main statutes that are relevant to rights to light are:

- ❖ Prescription Act 1832
- ❖ Rights of Light Act 1959.

There exists a large body of rights to light case law. The selection of case reports listed below includes the more recent and more important decisions.

- ❖ Allen and Another v Greenwood and Another [1975] 1 All ER 819 6, 35-6
- ❖ Bowring Services Ltd v Scottish Widows Fund & Life Assurance Society [1995] 16 EG, 206.
- ❖ Carr Saunders v Dick McNeil Associates Ltd and Others [1986] 1 WLR 992 37, 43
- ❖ Charles Semon & Co v Bradford Corporation [1922] 2 Ch 737
- ❖ Colls v Home & Colonial Stores [1904] AC 179 4, 9, 10, 29, 31-2, 34
- ❖ Dance v Triplow and Another [1992] 17 EG, 103
- ❖ Deakins v Hookings [1994] 14 EG, 133
- ❖ Ecclesiastical Commissioners for England v Kino [1880] 14 ChD 213 40.
- ❖ Fishenden v Higgs and Hill Ltd [1935] 153 LT 128 33
- ❖ Lyme Valley Squash Club Ltd v Newcastle under Lyme Borough Council and Another [1985] 2 All ER 405 28-9
- ❖ Marine and General Mutual Life Assurance Society v St. James Real Estate Co Limited [1991] 2 EGLR 178
- ❖ Ough v King [1967] 3 All ER 859 34
- ❖ Price v Hilditch [1930] I Ch 500 5, 37, 43
- ❖ Pugh and Another v Howels and Another [1984] 48 P CR 298 36-7
- ❖ Scott v Pape [1886] 31 ChD 554 6, 80
- ❖ Sheffield Masonic Hall Co v Sheffield Corporation [1932] 2 Ch 17 43
- ❖ Shelfer v City of London Electric Lighting Co [1895] 1 Ch 287 31
- ❖ Wheeldon v Burrows [1879] 12 ChD 31 78
- ❖ Wrotham Park Estate Co v Parkside Homes Limited [1973] ChD 321

Legislation

Daylight and sunlight amenity

As greater emphasis is placed on environmental issues, local planning authorities are increasingly concerning themselves with the effect of developments on the daylight and sunlight enjoyed by neighbouring properties. The local authority's Unitary Development Plan or Local Plan should always be consulted, as it will give an indication of what they expect in this regard.

The Building Research Establishment (BRE) published Report 209 in 1991 titled 'Site layout planning for daylight and sunlight: A guide to good practice' written by P J Littlefair. It was intended to give guidance on how to ensure good daylight and sunlight to proposed new development through good design, while avoiding detrimentally affecting neighbours' daylight and sunlight amenity. It sets out various tests that can be undertaken to establish if a problem is likely to be created.

The BRE report was not intended to be mandatory. However, it is not uncommon for planning authorities to require a developer to submit a daylighting and sunlighting study in support of a planning application and some may expect compliance with the recommendations given in the guidelines.

Daylight

The BRE report states that if any part of a new building or extension, measured in a vertical section perpendicular to a main window wall of an existing building, from the centre of the lowest window, subtends an angle of more than 25° to the horizontal, then the diffuse daylighting of the existing building may be adversely affected. This will be the case if either:

- ❖ the vertical sky component measured at the centre of an existing main window is less than 27%, and less than 0.8 times its former value; or
- ❖ the area of the working plane in a room which can receive direct skylight is reduced to less than 0.8 times its former value.

In such circumstances the occupants of the existing building will notice the reduction in the amount of skylight and more of the room will appear poorly lit.

Sunlight

The BRE report advises that new development should take care to safeguard access to sunlight for existing dwellings and any non-domestic buildings where there is a particular requirement for sunlight. If a living room of an existing dwelling has a main window facing within 90° of due south, and any part of a new development subtends an angle of more than 25° to the horizontal measured from a point 2 metres above ground in a vertical section, perpendicular to the window, then the sunlighting of the existing dwelling may be adversely affected.

The sunlight amenity of the existing dwelling will be considered to be adversely affected if the window reference point:

- ❖ receives less than one quarter of annual probable sunlight hours, and/or less than 5% of annual probable sunlight hours during the winter months between 21 September and 21 March, and the available sunlight hours in either period is reduced to less than 0.8 times its former value.

Analysis

The BRE report sets out the methods by which the effect of a development

Legislation

on existing neighbouring buildings may be calculated. If the local planning authority requires an assessment to be undertaken and submitted in support of a planning application, it is usual for the analysis to follow the tests in BRE Report 209.

The target values recommended in BRE Report 209 are quite demanding and can be difficult to achieve, particularly in dense urban environments. Care may need to be taken when applying the guidance and interpreting the results of analyses.

Party wall procedure

The Party Wall etc Act 1996 came into force throughout England and Wales on 1 July 1997. All previous local enactments, including the London Building Acts (Amendment) Act 1939 Part VI and the Bristol Improvement Act 1847, have now been repealed.

The general principle of the Act is to enable an owner to undertake certain specific works on, or adjacent to, adjoining properties while giving protection to potentially affected neighbours. As suggested by the word 'etc' in its title, the Act relates not just to party walls.

The parties

The owner of the property where the work is to be undertaken is the "building owner". The owner of the adjoining property is the "adjoining owner". There can be many adjoining owners including freeholder, leaseholder and anyone with an interest greater than from year to year. Under the Act, the word 'owner' can also include people with a contract to purchase or an agreement for lease; this arrangement allows a prospective owner to serve notice, and even commence work, before completion of the contract.

Party structures

A party wall is one that stands on the land of two owners, by more than its footings, or one which separates buildings of different owners. In the first case, the whole wall is a party wall whereas in the second, it is only a party wall for the extent to which the two buildings are using it and the whole of the rest of the wall belongs to the person on whose land it stands. Other types of party structure are party fence walls (ie shared garden walls) and party floors (eg separating different flats).

In the case of a party wall, each owner owns the part of the wall that stands on their own land, but also has rights over the remainder of the wall. The Act allows owners to treat the whole of the party wall as if it were their own, and debars them from dealing with their half on its own without informing their neighbours.

Before exercising any of the rights bestowed upon them, owners must follow the procedures set down in the Party Wall etc Act 1996. If an owner wishes to underpin, raise, cut into, thicken or demolish and rebuild a party wall, they must give notice of their intention to do so. In the event that the adjoining owner disagrees, each party must appoint a surveyor and a formal agreement known as a party wall award must be entered into.

An adjoining owner may respond to a party structure notice by serving a counter notice requiring certain specified works to be undertaken to the party wall for his future purposes. However the notice must be served within one month of the original notice and the adjoining owner will be responsible for the cost of the additional work.

Legislation

Excavations adjacent to structures

The Party Wall etc Act 1996 requires notice to be served on adjoining owners of certain intended excavations within 3 metres or 6 metres of adjacent buildings or structures.

Refer to figure below for details of these notifiable excavations.

3 metre notice

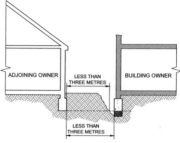

6 metre notice

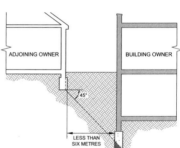

The excavations could be for any purpose, not just for a building or its foundations.

New walls at boundaries

Where it is proposed to build at the 'line of junction' (ie the boundary) notice may need to be given to the adjoining owners. This will be the case where the boundary is not already built upon, or is built upon only to the extent of a boundary wall (not being a party fence wall or external wall of a building). In return, a right to construct projecting mass concrete footings work may be claimed.

General rights and obligations

The Party Wall etc Act 1996 grants various general rights and obligations in relation to work undertaken in pursuance of the Act. The key ones are as follows:-

- ❖ obligation to serve notice on adjoining owners before undertaking the work;
- ❖ obligation to execute the work in compliance with other statutory requirements and in accordance with details agreed either by the parties or by appointed party wall surveyors in an Award;

Legislation

- right of access to enter and remain on adjoining owners' land so far as is reasonably necessary to facilitate the work;
- duty not to cause unnecessary inconvenience;
- duty to compensate adjoining owners for loss or damage arising as a consequence of the work; and
- obligation to make good damage caused by work to party structures.

Agreeing the works

The building owner and adjoining owner may agree the details of the works between them. Where disagreements occur, either deemed or actual, the Act provides a mechanism for resolving them through the appointment of party wall surveyors. The appointed surveyors have a duty to settle the matter by making an award and they must act expediently and impartially, having regard to the interests and rights of both parties.

Dos and don'ts

- Do consider whether proposed excavations and building works require the serving of notices under the Act.
- Do allow adequate time in the project programme for identifying and resolving all party wall issues. Time periods stated in the Act are statutory minimums, and often longer periods should be allowed.
- Do establish which party will act as the 'building owner' and verify his legal interest in the property.
- Do remember that if the building owner parts with his interest in the property part-way through proceedings, matters will have to start afresh.
- Do remember that the interests of all relevant adjoining owners will have to be identified with certainty and each owner notified separately.
- Do remember that each adjoining owner has a right to disagree with the proposals and to appoint a surveyor.
- Don't let unreasonable or unresponsive adjoining owners or surveyors hinder progress: use the mechanisms in the Act to force matters along.
- Do make allowance within the cost plan for the reasonable fees of the adjoining owners' surveyors and any necessary sub-consultants (typically structural engineers).
- Do ensure that the design team and contractors co-operate and produce all necessary drawings, method statements, calculations, etc, in good time.
- Do ensure that the proposed time and manner of carrying out the work is reasonable and that no unnecessary inconvenience will be caused.
- Do consider making provision within tender documents for restrictions on working hours and/or methods of working for noisy elements of work falling within the scope of the Act.
- Do ensure that the procedures required by the Act are followed meticulously; otherwise, notices and awards could be invalidated.
- Do remember that the party wall surveyors are administering the Act impartially and not representing clients.
- Do ensure that awards are agreed prior to starting the relevant work.
- Don't confuse common law and rights to light matters as being part of party wall procedures.

Legislation

- ❖ Do establish a line of communication between adjoining owners and contractors for dealing with day-to-day issues of noise, dust, and so on.
- ❖ Do ensure that any variations in the agreed works are agreed between the parties, or by their appointed surveyors.
- ❖ Finally, do ensure that all parties fulfil their obligations.

References

'Party Wall Legislation and Procedure' RICS Guidance Note (5th edition), Surveyors Publications, 2002.

'The Party Wall Act Explained - A Commentary on The Party Wall etc Act 1996' (The Green Book) - Pyramus and Thisbe Club, 2nd edition, 1997.

'Party Walls - The New Law' - S Bickford-Smith and C Sydenham, 2nd edition, Jordans, 2003.

'A Practical Manual for Party Wall Surveyors' - J Anstey, RICS Books, 2000.

'Party Walls and What to Do with Them' - J Anstey, 5th edition, RICS Books, 1998.

'An Introduction to the Party Wall etc. Act 1996', - J Anstey and V Vegoda, Lark Productions, 1997'The Party Wall Casebook' - P Chynoweth, Blackwell Science, 2003.

'Party Wall etc Act 1996' audio cassette - Owlion.

'Party Walls - Best Practice Roadshow' audio CD - Owlion, 2002.

'Party Wall etc. Act, 1996: explanatory booklet' - Office of the Deputy Prime Minister (ODPM).

Useful websites

Party Wall etc. Act, 1996 - HMSO
www.legislation.hmso.gov.uk/acts/acts1996/1996040.htm

Royal Institution of Chartered Surveyors
www.rics.org/public/party_walls/

Pyramus and Thisbe Club
www.partywalls.org.uk

Office of The Deputy Prime Minister
www.safety.odpm.gov.uk/bregs

Party Wall and Rights to Light Discussion Forum
www.partywallforum.co.uk

Access agreements

Access to Neighbouring Land Act 1992

The Access to Neighbouring Land Act 1992 enables a court to grant an order for access to land where such access is required to enable the execution of repair or maintenance work and where access has been refused by the neighbour.

The Act contains provisions for the preparation of schedules of condition and for overseeing of the work by surveyors. Surveyors can also be called upon to provide evidence where claims for damage under the Act are made.

Often the knowledge that rights exist under the Act will prompt neighbours to be accommodating, but, even then, it is still sensible to enter into an informal access agreement in the form of a licence with accompanying schedule of condition.

Legislation

Rights of way and fire escape agreements

More intense use of an existing right of way often requires re-negotiation of the right altogether. One type of right commonly encountered relates to fire escape routes benefiting properties that adjoin a development. The rights are often disrupted by large-scale redevelopment and negotiations for both temporary escape rights during the redevelopment and for revised escape rights over the new permanent development will be required.

Oversailing cranes and encroaching scaffolds

Most large developments necessitate the use of at least one crane. There are benefits to the developer or contractor of using fixed-jib tower cranes, as opposed to luffing-jib or folding-jib cranes, such as reduced cost and increased speed and load capacity. However, where the jib of a crane will oversail adjoining land, the agreement of the adjoining land owner, and any other party with an interest in the air space above it, will be needed.

Without such agreement the developer will be committing a trespass on every oversailing occasion.

The law on this matter comes from the tort of trespass. The position was clarified in the case of Anchor Brewhouse Developments Ltd and others v Berkeley House (Docklands Developments) Ltd [1987], in which the plaintiffs sought, and were granted, injunctive relief in relation to an unauthorised oversailing crane.

While some developers manage to place oversailing risks on their contractors, when work is abundant such contractors will often not accept such risks or, if they do, their tender prices are significantly enhanced.

Similar trespass issues often arise in relation to temporary independent scaffolding, although, wherever possible, rights granted by the Party Wall etc Act 1996 and the Access to Neighbouring Land Act 1992 should be exercised.

Developers and/or their contractors should give consideration at an early stage to the issue of access onto or over adjoining land or the airspace above it. Neighbours should be approached in advance for consent and the terms of such consent will be the subject of negotiation by the parties. It is sensible to enter into an access agreement or licence and the developer may be expected to pay a sum of money in consideration or grant reciprocal oversailing rights. Typically such agreements will provide for indemnities, insurance arrangements, schedules of condition, payment of fees and costs and so on.

Construction noise and vibration

The law relating to construction noise and vibration is controlled by the common law of nuisance and a number of statutes, in particular the Control of Pollution Act 1974.

Nuisance

A private nuisance is "an unreasonable interference with a person's use or enjoyment of the land". If a neighbour can demonstrate that a nuisance exists and that he has suffered substantial damage as a result, then he may be successful in bringing an action against the parties creating the nuisance.

The remedies available are injunctions and/or damages. The courts have, however, traditionally regarded demolition and construction sites as a special case as far as noise is concerned. It would appear that as long as the works are carried out with proper skill and care, and all reasonable

Legislation

precautions are taken to minimise disturbance to neighbouring occupiers, no action in nuisance will arise.

Developers and contractors have to ensure that they take "reasonable precautions". They must be aware of the definition of "best practicable means" in the Environmental Protection Act 1990. This includes considering local conditions and circumstances, the current state of technical knowledge and financial implications.

A leading case on construction noise and nuisance is Andreae v Selfridge [1938] 3 All ER 264 where the judge held that provided demolition and building operations are "reasonably carried on, and all proper and reasonable steps are taken to ensure that no undue inconvenience is caused to neighbours, whether from noise, dust or other reasons, the neighbours must put up with it".

Control of Pollution Act 1974

The Control of Pollution Act 1974 deals, in part, with the control of noise on construction sites. Section 60 empowers a local authority to serve notice imposing certain limitations. These limitations include specifying the hours of work, permitted noise levels and the particular plant and machinery that may be used. The recipient may appeal against the notice within 21 days. Contravention of the notice is an offence under the Act. Section 61 provides an opportunity for a developer or contractor to apply to the local authority in advance of the works and seek agreement to such matters as the method of carrying out the work and the steps that will be taken to minimise noise. The local authority is not obliged to give its consent and, even if it does, it may attach conditions to it. It may also change the conditions if it sees fit.

The local authority will often specify that "best practicable means" are used. The developer/contractor will be expected to adopt the quietest viable method of working within reasonable cost limits. The local authority may also require continuous noise and vibration monitoring and regular liaison meetings with neighbours.

Architectural and design criteria

Design

- Basic design data — 110
- Building types — 111
- Specifications — 115
- Different specifications for different contracts — 115
- Specification writing — 115
- Site archaeology — 116
- Coordinating project information — 118

Design

Basic design data

Internal circulation

Space allowances: (minimum areas per person)

Building type	m²	ft²
Offices (excluding cores)	9.3	100
Retail	4.6-7.0	50-75
Factories	7.0	75
Restaurants	0.9-1.1	10-12

Lighting requirements

Circulation: 150 Lux
Casual work: 200 Lux
Routine work: 300 Lux
Offices: 350 Lux - 500 Lux
Drawing office: 750 Lux
Fine work: 1000 Lux
Very fine work: 1500 Lux
Minute work: 3500 Lux

Noise and acoustics requirements

Auditoria: 20-30 dB (A) Leq
Bedroom: 30-35 dB (A) Leq
Small office: 40-45 dB (A) Leq
Large office: 45-50 dB (A) Leq
Light industrial: 50-55 dB (A) Leq

Design values For internal environmental temperatures

Flats, residences and hotels
Living rooms — 21°C
Bedrooms — 18°C
Bathrooms — 22°C
Staircases and circulation — 16°C

Offices
General — 20°C
Private — 20°C
Circulation — 16°C

Retail
Small — 18°C
Large — 18°C
Department stores — 18°C
Fitting rooms — 21°C
Store rooms — 15°C

Restaurants — 18°C

Factories
Sedentary work — 19°C
Light work — 16°C
Heavy work — 13°C

Hotels
Bedrooms — 22°C
Public rooms — 21°C
Circulation — 18°C

Design

External circulation

Car parking	Standard bay 2.4 x 4.8m	Allow 6.1m for head-on parking. Area per car 18.8m^2
Ramps	Car parking garages	10%
	With transition ramps at half the ramp gradient for 2.4m at each end	15%
	Pedestrian	10%
Carriageway widths	One-way four lanes	14.6m
	One-way two lanes	7.3m
	Two-way two lanes	max 7.3m, min 6.0m
Vehicle sizes	Cars max length	5.7m
	Cars min length	3.05m
	Lorries max length	18.0m
	Vans max length	6.0m
	Standard refuse lorry length	7.4m
	Standard fire appliance length	8.0m

Building types

Housing

Typical approximate housing equivalent densities:

50 houses per acre = 124 houses per hectare
40 houses per acre = 99 houses per hectare
30 houses per acre = 74 houses per hectare
20 houses per acre = 49 houses per hectare
10 houses per acre = 25 houses per hectare

Recommended planning grid – 300mm
Recommended minimum floor to ceiling – 2300mm
Minimum floor to floor – 2600mm

Recommended areas for number of persons per dwelling (m^2) are identified in the folllowing table.

Houses:	1	2	3	4	5	6	7
One storey	30	44.5	57	67	75.5	84	
Two storey				72	82	92.5	108
Three storey					94	98	112
Flats	30	44.5	57	70	79	86.5	
Maisonettes				72	82	92.5	

Parker Morris is the preferred standard for new buildings by local authorities and housing associations. Private housing is subject only to space requirements by the Public Health Acts.

Design

Hotels

Types
- city centre hotels
- motor hotels
- airport hotels
- resort hotels
- motels

Space allocation (m²/room gross)
- city centre hotel   45-65
- motor hotel   35-45
- resort hotel   40-55

Hotels under 77-80 rooms only viable as family run businesses.

Fire precautions
Travel distances for escape are dealt with in the guide to the Fire Precautions Act 1971.

Car parking
- resident guests: one space/bedroom
- conference facilities: one space/five seats

Required provision must be agreed with the local authority.

Offices

Methods of calculating areas
- For planning purposes, gross total area measured over external walls.
- For cost purposes, gross total area measured inside external walls.
- Nett areas are measured between inside walls and exclude core areas, ducts and staircases.

Definition of space
- very deep space: over 20m
- deep space: 11-19m
- medium deep space: 6-10m
- shallow space: 4-5m

Dimensional criteria
- planning grids: 900mm 1200mm 1500mm
- floor-to-floor heights: 2700mm-5100mm
- floor-to-ceiling heights: 2400mm-3000mm
- floor zone: 300mm-1200mm

Means of escape
- Maximum travel distance with escape in only one direction – 12.2m.
- Maximum travel distance with escape possible in alternative direction – 46m.
- Maximum distance between exits on a storey – 61m.

At least one fire fighting stair required with floor more than 18m above ground floor level.

Lavatory provision based on the Offices, Shops & Railway Premises Act 1963 Sections 9 & 10.

Design

Car parking
- ❖ Staff: one space for each 25m² of gross floor area.
- ❖ Visitors: 10% of parking provision required.

Required provision must be agreed with the local authority.

Recreational

The dimensions of a number of particular sports facilities are identified in the following table.

Outdoor sport facilities

Olympic standard swimming pools	length width constant depth	50m 21m 1.8m
Running tracks:	200m 300m 400m	86.79 x 61.78m 126.52 x 86.48m 170.91 x 133.00m
Association Football pitch size		100m x 60m
Rugby Union pitch size		144m x 69m
Hockey pitch size		91.5m x 54.9m
Tennis court size		10.97m x 23.77m

Indoor sport facilities

Netball court size		30.50m x 15.25m
Badminton court size		13.4m x 6.1m
Table tennis (international)		14.0m x 7.0m

An indoor space 32m x 26m can contain the following sports:	Two badminton courts, karate, fencing, table tennis, basketball, archery, volleyball, soccer, hockey.

Design

Industrial

Site coverage and floor loading standards for industrial property are identified in the following table.

Site coverage:	Plot ratio normally maximum of 1:1 including office content. Site coverage should not exceed 75% (normally approx. 50-60%)	
Car parking in factories	Staff:	1 car/50m² of gross floor area
	Visitors:	10% of staff parking
Car parking in warehouses:	Staff:	1 car/200m² of gross floor area.
	Lorry parking: Minimum standards for loading bays:	70m² for every 100m² gross floor space
		140m² for every 250m² gross floor space
		170m² for every 500m² gross floor space
		200m² for every 1000m² gross floor space
		300m² for every 2000m² gross floor space
		50m² for every additional 1000m² gross floor space
Factory building types:	Light duty industrial:	Spans min 9m max 12m Ht. to eaves 4.5m floor loading 16kN/m²
	Medium duty industrial:	Spans 12m-18m Ht. to eaves 6.5m Floor loading 25kN/m²
	Heavy duty industrial:	Spans 12m-20m Ht. to eaves 7m-12m Floor loading 15-30kN/m²
Warehouses:	General purpose:	Spans 12-18m Ht. to eaves 8m Floor loading 25kN/m²
	Intermediate high bay:	Spans 12-20m Ht. to eaves 14m Floor loading 50kN/m²
	High bay:	Ht. to eaves 30m Floor loading 60kN/m²

Design

Specifications

The specification is the hub around which the various documents forming a modern day building contract are assembled.

The specification will incorporate a complex array of information, which will include design, testing, procedural, visual and quality control information.

The parties will seek to have fully prescribed the product to which they are to sign up under a building contract. The standard forms of building contract generally recognise the specification as a contract document, but there are exceptions such as the JCT98 Standard Form with Quantities, which does not and care must be taken to ensure that the wording is amended to include the specification.

The specification as a building contract document will typically be supported by the preliminaries and either bills of quantities or contract sum analysis. It will schedule the drawings to which it refers, and which should form the contract drawings.

Different specifications for different contracts

Design specification

This form of specification will be used where the author wishes to prescribe the project in as much detail as possible, no imagination is required as to the clients' requirements. Products will be named and the assembly of the building will be described and supported with drawn information on a 'do as I say' basis.

The form of contract employed here will be a traditional one which will not carry a design responsibility for the contractor.

Performance specification

A performance specification sets out design parameters to which contractors are invited to submit their proposals. This can be a procurement route used where the design team have little idea as to products or visual requirements.

The client will not need to advance the design as far before inviting tenders from interested contractors. This is a more economic approach since pre-contract design costs are usually less at the early stages of procurement.

The specification for a design and build contract will initially form the employers requirements latterly supported by the contractors proposals once agreed with the selected contractor. This approach has the benefit of the contractor bringing the supply chain knowledge to the client at an early stage while still in competition.

The parameters of design so described in a performance specification can either be laid down loosely or brought in very tight according to the degree of control the client wishes to retain.

Specification writing

Construction consultants have established standardised data at their disposal to assist and form the basis to prepare their documents. Online specification writers are available, the most widely used is National Building Specification (NBS) - www.nbsservices.co.uk. This is a subscription service.

Design

The NBS system stores a fully comprehensive set of specifications for most conceivable products and materials which they may adapt, add to or modify according to a particular project.

The NBS has recently published 'NBS Educator' which is designed to provide construction industry students, lecturers and professionals who need to understand contract documentation, such as specifications and schedules of work. The NBS Educator website includes a variety of functions to navigate which provides the user with suggestions and ideas as well a reference library for different aspects of specification writing.

Another specification tool that can be accessed online is Barbour Index which has recently launched its Specification Expert website which is at www.barbour-index.co.uk. The service is designed to create, edit and manage specification documents. The service makes writing specifications quick and easy, offering a comprehensive library of all major specification clauses. It is available in two versions: Building (Architectural and, Building Surveying) and Building Services which incorporates the National Engineering Specification.

Site archaeology

While the British Government, including associated Scottish Parliament, Welsh Assembly and Northern Irish Authorities, are consulting on changes to Heritage, Scheduled Ancient Monuments and Archaeological Protection Legislation including the production of Unified Lists covering all categories, no final decisions have yet been made. Accordingly, the following procedures remain currently in place. There are indications that both White and Green papers in these subjects will be issued sometime in 2004 but no dates are yet confirmed at the time of writing.

The 'Little Oxford Dictionary' definition of archaeology is:

> Archaeology - the study of ancient peoples, especially by excavation of physical remains.

Not everyone will realise that in the construction world we have to deal with two types of archaeology:-

- ❖ The traditional excavation of what lies hidden below ground.
- ❖ The archaeology of standing buildings or above-ground archaeology.

How does archaeology affect construction?

Archaeology affects construction primarily through the way that the planning process limits and controls development. Both above and below ground archaeology can easily be applicable to the same site.

The present key government reference document is Planning Policy Guidance Document 16 published in November 1990 (PPG16). The government is now consulting for the replacement for PPG16 with the first of the new Planning Policy Statements (PPS). In this case the Planning Policy Statement on Planning for the Historic Environment.

There is as yet no formal indication as to when PPG16 is to be replaced by the PPS.

PPG16 sets out the secretary of state's policy on archaeological remains on land, and how they should be preserved or recorded both in an urban setting and in the countryside.

Design

Planning permission is required for works of development – the carrying out of building, engineering, mining or other operations in, on over or under land or the making of any material change in the use of any buildings or other land.

The first port of call for any planning application is the local planning office and the published documentation in either the Development Plan or the Unitary Development Plan.

If the Local Development Plan and/or the Local Planning Office identifies a site as one with scheduled ancient monuments or locations where archaeological remains are believed to exist then the earliest possible consultation with the county archaeologists or their London equivalents at English Heritage is vital.

This consultation will identify the archaeological sensitivity of the site.

Following discussions with the county archaeologists the next step for the developer is to have researched exactly what is or might be there and how it will or could impinge on his proposed development.

This is a cost for the developer and will have to be paid for by the developer in the same way as any other site investigation exercise is paid for prior to a development being undertaken.

Options for archaeological investigations are potentially twofold.

Desktop studies

These draw on the archaeological records of the site itself and those of neighbouring sites. They are relatively quick and easy. Crucially they must be done by a trained archaeologist. The handbook of the Institute of Field Archaeologists will help in locating the right archaeologist or archaeological unit to undertake the study. Contact can be made on admin@archaeologists.net or by telephone on 01189 316446.

Discussions may however indicate that the next level of investigation is necessary.

Field evaluation

This is not a full archaeological elevation or 'dig' but a ground survey and small scale trial trenching.

The results from the desktop study and/or the field evaluation will inform both the developer and the county archaeologist and the local planning officers, especially the conservation officer, as to the importance of the site in archaeological terms.

Model agreements between developers and the appropriate archaeological body regulating archaeological site investigations and excavations can be obtained from the British Property Federation, publications@bpf.org.uk or telephone 020 7828 0111.

Statutory protection for archaeological sites

Where Scheduled Ancient Monuments have been identified under the Ancient Monuments and Archaeological Areas Act 1979, these are sites of national importance and as such rank as the equivalent of Grade I or Grade II* buildings of special architectural or historic interest.

The protection of Scheduled Ancient Monuments is based on the requirement that scheduled monument consent has to be obtained from the secretary of state (now of the Department of the Environment and the Regions, previously the Department of the Environment).

Application has to be made to the secretary of state for any works that would have the effect of:

Design

> Demolishing, destroying, damaging, removing, repairing, altering, adding to, flooding or covering up the monument.
>
> If you find a scheduled ancient monument standing on or laying below your site take great care. Ensure your professional advisors fully appreciate the implications before application is made. If in doubt, enter into the earliest possible discussions with the relevant heritage and conservation authorities covering the site. These will include the local authorities conservation officer, the county archaeologist, and the Inspectors of Ancient Monuments at English Heritage.

Above ground archaeology

A Scheduled Ancient Monument is usually, although not exclusively, an unoccupied structure and can be either remains below ground and/or a standing structure above ground.

Above ground Scheduled Ancient Monuments will require Scheduled Ancient Monuments Consent for works to the structure. To fully understand what is important in this structure in understanding 'ancient peoples by the study of their physical remains' was borne the concept of above ground archaeology. This can involve the physical digging in to the structure above ground. This can equally involve the 'desktop study' approach of bringing together the known archive or history of the structure to fully understand what is important in the remains and what is less important.

This is because as structures are added to and/or changed over time there may well be pieces or parts of what is now extant that can well be removed or modified or changed and benefit the scheduled monument.

The final use of archaeology in standing structures is the extrapolation of the 'desktop study' for above ground scheduled ancient monuments into the 'Conservation Plan' for all types of listed historic buildings.

Conservation Plans are increasingly required by the Heritage and Listed Building authorities, particularly for the more important Grade I and Grade II* buildings as listed as being of Special Architectural and Historic Interest. This is as part of the application process for Listed Building Consent for works that will affect the special architectural or historic interest of these Listed Buildings.

Conservation plans have a number of practical and financial advantages for the developer/renovator or historic buildings. The key ones are:

(a) Certainty for the design team in formulating in what can or cannot be done to the building and site before they formulate details for the Listed Building Consent and Planning Permission applications.

(b) Certainty for the developer/renovator that the scheme will not be stopped in its tracks by the Heritage authorities discovering that the building contains important previously unknown historic elements in the structure that precludes all or part of the development going forward. The worst case being the stopping of the scheme or at least making it less viable or completely unviable.

Coordinating project information

Some of the characteristics of the construction industry make the quantity and timing of information flows on construction projects difficult to administer. These include the following:-

- ❖ Large numbers of participating companies collaborating on projects of relatively short duration.
- ❖ Lack of control over the location and working conditions of construction sites.
- ❖ Comparatively low levels of management support.

Design

Despite this, the use of electronic systems for information management including electronic tendering and project management systems can (when carefully implemented and managed) have several benefits including:

- ❖ reduced paper trails on labour intensive tasks;
- ❖ reduced costs and time savings;
- ❖ improved management of information;
- ❖ a platform to stimulate innovation; and
- ❖ the support of a green agenda.

Relevant legislation/information

The British Standards Institution's section for delivering information solutions to customers (BSI-DISC) sets out five principles which should be the starting point for coordinating information in 'DISC PD 0010:1997. The principles of good practice for information management' are as follows:-

- ❖ Recognise and understand all types of information;
- ❖ Understand the legal issues and execute 'duty of care' responsibilities;
- ❖ Identify and specify business processes and procedures;
- ❖ Identify enabling technologies to support business processes and procedures; and
- ❖ Monitor and audit business processes and procedures.

There is a need at the inception of any construction project to agree the principles in order to effectively manage communication and the flow of information. The above principles are relevant whether a web-based system is being used on a large prime contract project or a paper-based system on a smaller traditional project. The principles should be agreed with all members of the project team and documented with their conditions of appointment.

The DTI, Building Centre Trust and Phrontis Ltd have, in partnership, produced a Construction Best Practice report titled 'Effective Integration of IT in Construction' (October 2001). It highlights three key sectors in which e-business can offer improved means of coordinating project information including; specification, tendering and the supply chain.

- ❖ The internet is currently providing many opportunities for materials and equipment firms to provide information and communicate more directly with specifiers.
- ❖ In connection with e-tendering, the Office of Government Commerce (OGC) has been using a pilot system for several contracts and it is the government's intention to introduce e-tendering in central civil government by the end of 2002.
- ❖ To achieve success for e-business supply chains is difficult.

The nature of relationships is generally transitory, driven by project type and locality. Contractors may be put off investing in this area because of the threat of reduced margins.

The report concludes that e-business offers the construction industry a powerful array of tools to communicate and collaborate and should go some way to help overcome many of the endemic problems of the industry.

'A Code of Procedure for Construction Industry on Production Information' has been issued by the construction industry (issued by the DTI/CCPI September 2002). The aim of the code is to provide authoritative practical guidance on the preparation of drawings, specifications and schedules of work, making optimum use of widely adopted systems.

'BS7799 Information Security Management' provides guidance on security of information, which is of particular relevance within partnering relationships. A continuing push to use IT to conduct business electronically and globally requires a high degree of trust between customers and suppliers and between trading partners. The aim of BS7799 is to give guidance on internal management systems, to improve confidence between trading partners and to ensure compliance with the 1998 Data Protection Act.

A report entitled 'Benchmarking Electronic Service Delivery' by the central IT unit of the Cabinet Office compares the UK's progress on the development of electronic government services with that of other major economic nations. The UK has set more ambitious targets than most of its competitors, with an aim to provide 100% of all government services electronically by 2005.

Examples of project document management systems

The following list summarises a small sample of project document management systems currently available on the World Wide Web:

Asite - www.asite.com

Cadweb - www.cadweb.co.uk

Citadon - www.citadon.com

Meridian - www.mps.com

Recommendations

The recommendations for overcoming barriers to IT integration for coordinating project information, as summarised from 'Effective Integration of IT in Construction' include the following:-

- ❖ Adopt a suitable legal framework for using electronic communications.
- ❖ Appoint a leader with IT skills early in a project.
- ❖ Develop an information management strategy at project inception.
- ❖ Consider integration skills as a factor in the selection of the professional team.
- ❖ Ensure each team participant's in-house information systems are robust.
- ❖ Consider using a collaborative approach.

In conclusion, a complete understanding of the project, the environment and the people involved is necessary to effectively coordinate project information.

Structural and civil engineering design

- The design process   122
- Soils and foundation design   122
- Soil survey   123
- Design loadings for buildings   126
- Floors in industrial and retail buildings   130

Design

The design process

The Institution of Structural Engineers defines structural engineering as:

> The science and art of designing and making, with economy and elegance, buildings, bridges, frameworks and similar structures so that they can safely resist the forces to which they may be subjected.

Generally, engineers use limit state design methods. These model the way in which loadings are applied to a building over the course of its life.

Limit state design utilises differing probability factors for separate types of loads and material stresses to reflect the degree of certainty in assumptions made about each. For instance, strength is checked at the ultimate limit state where the building or element is on the point of collapse.

Deflection and vibration are checked at the serviceability limit state, at which point the building is considered to be unserviceable, even though it will not collapse.

Today, design in concrete, steel and masonry is usually carried out using limit state methods, with timber design scheduled to change over with the imminent arrival of the new Eurocode. Foundation design, involving the much more unpredictable material of the soil, has still eluded the process.

An assessment of ground conditions is usually essential to ensure the proper performance of the superstructure.

Soils and foundation design

For very light residential building, a visual assessment of the soils with a minimum of physical testing may be sufficient, if local knowledge of the area is good. For larger buildings, a proper geotechnical investigation is usually recommended. This allows the measurement of relevant soils (geotechnical) data on which the foundation design will be based. Accurate testing of samples of soil obtained from boreholes usually leads to economies in foundation design. Samples are usually tested for consolidation, compressibility (both of these affect settlement) and shear strengths. In certain circumstances testing for chemicals, which could affect the structure, may be included.

Where soil conditions are good, and building loads are relatively light, a simple spread foundation is usually sufficient. This will be in the form either of simple strips or pad bases of non-reinforced concrete. As loads increase and soil conditions become more complex, foundations may need to be reinforced to allow loads to be spread sufficiently. Heavy building loads and poor ground often require piled foundations to take the foundation loads into suitable (deeper) soils.

Other features, such as trees (in shrinkable soils), groundwater, and basements, all require special consideration in foundation design.

The relevant British Standards to which reference should be made are BS 8004 Code of Practice for Foundations and BS 5930 Code of Practice for Site Investigation.

For preliminary design purposes only, the following table gives typical allowable bearing values under static loading for various types of soils and may be used to size strip or pad foundations.

Category	Soil type	Presumed allowable bearing pressure (kN/m^2)
Un-weathered rocks	Strong igneous rocks,	10,000
	Strong sandstones and limestones	4,000
	Strong shales	2,000
Non-cohesive soils	Dense gravel	600
	Medium dense gravel or sand and gravel	200-500
	Loose gravel or loose sand and gravel	up to 200
	Loose sand	up to 100
Cohesive soils	Very stiff clay	300-600
	Stiff clay	150-300
	Firm clay	75-150
	Soft clay	75

Soil survey

Purpose

Soil surveys may be required to determine either, or both, of the following:-

- ❖ Environmental contamination.
- ❖ Geotechnical properties.

Environmental

An intrusive investigation may be required to establish the nature of any sub-surface contaminated soil and/or groundwater. Site investigations are based upon site-specific factors such as the presence of likely contaminants, the anticipated sub-surface geology and the location of the site. They are often referred to as Phase 2 investigations where they follow a non-intrusive Phase 1 audit. Samples of soil and groundwater are taken for laboratory analysis. The results are interpreted and a risk assessment undertaken for any contamination identified.

Geotechnical

The purpose of a geotechnical site investigation provides sufficient geotechnical information to enable design of foundations. Investigations will be designed specifically to suit the anticipated geology and development proposals. A geotechnical site investigation is intrusive and may include in-situ and laboratory tests.

A geotechnical site investigation is often procured with an environmental site investigation given a significant overlap with the fieldwork requirements.

Design

Fieldwork

The following methods of soil and groundwater sampling are common to both environmental and geotechnical investigations.

Trial pits

Extremely valuable if the depth of investigation is less than about 3 metres. Trial pits allow a detailed examination of the ground conditions in-situ with some indication of stability and groundwater conditions. Trial pitting is a relatively fast and efficient means of exploring sub-surface conditions.

Auger holes

Auger holes are normally made by hand-turning a very light auger into the ground, or by using light power-auger equipment. However, while auger holes are effective in preliminary investigations they do not provide the depth or range of sampling and in-situ testing provided by conventional site investigation boreholes.

Window samplers

A window sampler is a steel tube, usually about 1 metre long, with a series of windows along the tube through which to view disturbed soil conditions or extract samples. A lightweight percussion hammer drives the sampler into the ground, which is then extracted with jacks. A depth of approximately 10 metres can be achieved, depending on soil conditions, using a sequence of progressively smaller diameter samplers.

Boreholes

Light percussion drilling (shell and auger) is the most commonly used method in the UK. Samples are often unsuitable for soil description on the basis that the drilling process will have changed the strength of the soil by remoulding it and increased the moisture content of the soil through lubrication. For this reason U100 tube samples are often taken from boreholes at regular depths by driving a small diameter tube into the soil at the base of the borehole, thus extracting an undisturbed sample for laboratory analysis. Lubrication may be required if the sub-surface is not cohesive making it unsuitable for some types of environmental site investigations.

Geotechnical in-situ testing

In-situ testing is typically undertaken on cohesionless soils where the strength characteristics may only be established on undisturbed in-situ samples.

Cone Penetration Test (CPT)

This technique involves hydraulically pushing a 10 or 15cm^2 cone into the ground at a standard rate of penetration and measuring the penetration resistance. The equipment necessary for this type of investigation is housed in a large truck. The results from a CPT can be very valuable providing a soil profile, estimates of the soil types encountered including consistency and density, and evidence of the presence of voids beneath the site.

Dynamic probing

Much less sophisticated than CPT, dynamic probing involves driving a steel rod into the ground by using repeated blows of a hammer of a specified mass falling through a fixed distance. The number of blows required for each 100mm is recorded and plotted as a depth versus blow-count log. However, the information given by dynamic probing is very restricted and is difficult to interpret. Further investigations are usually necessary to supplement this test.

The Standard Penetration Test (SPT)

This test involves driving an open-drive sampler into the bottom of a borehole with repeated blows of a hammer falling a predetermined distance. The number of blows necessary to drive the sampler six increments of 75mm are counted and recorded, giving the penetration resistance value. The test is undertaken on cohesionless soils such as gravels.

Other methods of testing and sampling are used depending on the soil type and conditions on site.

Geotechnical laboratory testing

Laboratory testing is typically undertaken on cohesive soils ie clays, where an undisturbed sample may be retrieved from the field. Such testing is not appropriate to cohesionless soils such as gravels. Tests include:-

- ❖ Soil Classification: Particle Size Distribution, Plasticity Index, Moisture Content.
- ❖ Consolidation: To assess the settlement characteristics.
- ❖ Triaxial: To assess the compressive strength characteristics.

Geotechnical reporting

The results of the fieldwork and laboratory tests are assessed. For development purposes preliminary designs for foundations may be prepared. Sufficient information would be provided to enable design of building foundations. For diagnostic purposes the investigation would assist with identifying the cause of defects with a building superstructure allowing remedial solutions to be specified with confidence.

Environmental sampling and analysis

Soil sampling

Soil samples are required for chemical analysis to determine the presence of any contamination. Methods of soil sampling differ depending on the contaminants being tested. Usually disturbed samples are adequate for testing most chemicals. These can be obtained from the excavator bucket when trial pits are being used, from the cuttings from boreholes or from the window within a window sampler.

Samples are analysed in the laboratory for a wide range of 'base line' contaminant chemicals often supplemented by further specialist testing depending upon the type of contamination present.

Water sampling

Water samples can be obtained by inserting a standpipe into a completed borehole. Water samples will be taken after insertion for further chemical analysis. The well may also be used for the monitoring of gas. Installation can be permanent/semi-permanent to facilitate further sampling at a later date. Laboratory analysis is similar to soil samples.

Gas sampling

The presence of methane and carbon-dioxide may be established on sites which are landfills or close to neighbouring landfills. Monitoring wells constructed for water sampling may also be used to sample the presence of land gas. In-situ measurements of gas concentrations and flow are taken at ground level on the head of the monitoring well using an infrared gas analyser. Wells may be constructed to measure gas concentrations and flow at different depths.

A simpler broad-brush method of establishing the general presence of land gas is to undertake a shallow 'spike survey'. The presence and extent of contamination from a leaking underground petrol tank is a particularly common example.

Design

Environmental risk assessment

On completion of the fieldwork and laboratory testing a risk assessment is undertaken. Where contamination has been identified it is important to consider the following:-

- ❖ The client's requirements.
- ❖ The present use of the site and risk to occupants.
- ❖ The future intended use of the site and any considerations triggered by redevelopment.
- ❖ Future excavations for foundations or buried services.
- ❖ Contamination of groundwater particularly where abstracted for water supply.
- ❖ Contamination of nearby water courses.
- ❖ Contamination to or from neighbouring sites.

Recommendations and budget costs should be provided for remediation.

Design loadings for buildings

The table overleaf (Table 1 BS6399: Part 1) indicates minimum recommended imposed loads for use in the design of certain parts or types of building. For the expanded list and limitations of use, reference should be made to BS 6399: Parts 1 and 3: 1996.

Table 1. Minimum imposed floor loads			
Type of activity/ occupancy part of the building or structure	Examples of specific use	Uniformly distributed load kN/m^2	Concentrated load kN
A Domestic and residential activities (Also see catagery C)	All usages within self-contained dwelling units communal areas (including kitchens) in blocks of flats with limited use (See note 1) (For communal areas in other blocks of flats, see C3 and below)	1.5	1.4
	Bedrooms and dormitories except those in hotels and motels	1.5	1.8
	Bedrooms in hotels and motels Hospital wards Toilet areas	2.0	1.8
	Billiard rooms	2.0	1.8
	Communal kitchens except flats covered by note 1	3.0	4.5
	Balconies — Single dwelling units and communal areas in blocks of flats with limited use (See note 1)	1.5	1.4

Design

Table 1. Minimum imposed floor loads (continued)

Type of activity/occupancy part of the building or structure	Examples of specific use		Uniformly distributed load kN/m²	Concentrated load kN
A Domestic and residential activities (Also see category C) (continued)	Balconies (continued)	Guest houses residential clubs and communal areas in blocks of flats except as covered by note 1	Same as rooms to which they give access but with a minimum of 3.0	1.5/m run concentrated at the outer edge
		Hotels and motels	Same as rooms to which they give access but with a minimum of 4.0	1.5/m run concentrated at the outer edge
B Offices and work areas not covered elsewhere	Operating theatres, X-ray rooms, utility rooms		2.0	4.5
	Work rooms (light industrial) without storage		2.5	1.8
	Offices for general use		2.5	2.7
	Banking halls		3.0	2.7
	Kitchens, laundries, laboratories		3.0	4.5
	Rooms with mainframe computers or similar equipment		3.5	4.5
	Machinery halls, circulation spaces therein		4.0	4.5
	Projection rooms		5.0	To be determined for specific use
	Factories, workshops and similar buildings (general industrial)		5.0	4.5
	Foundries		20.0	To be determined for specific use
	Catwalks		-	1.0 at 1m centres
	Balconies		Same as rooms to which they give access but with a minimum of 4.0	1.5/m run concentrated at the outer edge
	Fly galleries		4.5 kN/m run distributed uniformly over width	-
	Ladders		-	1.5 rung load

Design

Table 1. Minimum imposed floor loads (continued)

Type of activity/occupancy part of the building or structure	Examples of specific use		Uniformity distributed load kN/m²	Concentrated load kN
C Areas where people may congregate	Public, institutional and communal dining rooms and lounges, cafes and restaurants (See note 2)		2.0	2.7
C1 Areas with tables	Reading rooms with no book storage		2.5	4.5
	Classrooms		3.0	2.7
C2 Areas with fixed seats	Assembly areas with fixed seating (See note 3)		4.0	3.6
	Places of worship		3.0	2.7
C3 Areas without obstacles for moving people	Corridors, hallways, aisles, stairs, landings etc. in institutional type buildings (not subject to crowds or wheeled vehicles), hostels, guest houses, residential clubs and communal areas in blocks of flats not covered by note 1. (For communal areas in blocks of flats covered by note 1, see A)	Corridors, hallways, aisles etc. (foot traffic only)	3.0	4.5
		Stairs and landings (foot traffic only)	3.0	4.5
	Corridors, hallways, aisles, stairs, landings etc. in all other buildings	Corridors, hallways, aisles etc (foot traffic only)	4.0	4.5
	Hotels and motels and institutional buildings	Corridors, hallways, aisles etc, subject to wheeled vehicles, trolleys etc.	5.0	4.5
		Stairs and landings (foot traffic only)	4.0	4.0
	Industrial walkways (light duty)		3.0	4.5
	Industrial walkways (general duty)		5.0	4.5
	Industrial walkways (heavy duty)		7.5	4.5
	Museum floors and art galleries for exhibition purposes		4.0	4.5

Design

Table 1. Minimum imposed floor loads (continued)

Type of activity/occupancy part of the building or structure	Examples of specific use	Uniformly distributed load kN/m²	Concentrated load kN
C3 Areas without obstacles for moving people (continued)	Balconies (except as specified in A)	Same as rooms to which they give access but with a minimum of 4.0	1.5/m run concentrated at the outer edge
	Fly galleries	4.5 kN/m run distributed uniformly over width	-
C4 Areas with possible physical activities (See clause 9)	Dance halls and studios, gymnasia, stages	5.0	3.6
	Drill halls and drill rooms	5.0	9.0
C5 Areas susceptible to overcrowding (See clause 9)	Assembly areas without fixed seating, concert halls, bars, places of worship and grandstands	5.0	3.6
	Stages in public assembly areas	7.5	4.5
D Shopping areas	Shop floors for the sale and display of merchandise	4.0	3.6
E Warehousing and storage areas. Areas subject to accumulation of goods. Areas for equipment and plant	General areas for static equipment not specified elsewhere (institutional and public buildings)	2.0	1.8
	Reading rooms with book storage, eg, libraries	4.0	4.5
	General storage other than those specified	2.4 for each metre of storage height	7.0
	File rooms, filing and storage space (offices)	5.0	4.5
	Stack rooms (books)	2.4 for each metre in storage height but with a minimum of 6.5	7.0
	Paper storage for printing plants and stationery stores	4.0 for each metre of storage height	9.0

Design

Table 1. Minimum imposed floor loads (continued)

Type of activity/ occupancy part of the building or structure	Examples of specific use	Uniformly distributed load kN/m²	Concentrated load kN
E Warehousing and storage areas. Areas subject to accumulation of goods. Areas for equipment and plant (continued)	Dense mobile stacking (books) on mobile trolleys, in public and institutional buildings	4.8 for each metre of storage height but with a minimum of 9.6	7.0
	Dense mobile stacking (books) on mobile trucks, in warehouses	4.8 for each metre of storage height but with a minimum of 15.0	7.0
	Cold storage	5.0 for each metre of storage height but with a minimum of 15.0	9.0
	Plant rooms, boiler rooms, fan rooms, etc, including weight of machinery	7.5	4.5
	Ladders	-	1.5 rung load
F	Parking for cars, light vans, etc. not exceeding 2500 kg gross mass, including garages, driveways and ramps	2.5	9.0
G	Vehicles exceeding 2500 kg. Driveways, ramps, repair workshops, footpaths with vehicle access, and car parking	To be determined for specific use	

NOTE 1. Communal areas in blocks of flats with limited use refers to blocks of flats not more than three storeys in height and with not more than four self-contained dwelling units per floor accessible from one staircase.

NOTE 2. Where these same areas may be subjected to loads due to physical activities or overcrowding, eg, a hotel dining room used as a dance floor, imposed loads should be based on occupancy C4 or C5 as appropriate. Reference should also be made to clause 9.

NOTE 3. Fixed seating is seating where its removal and the use of the space for other purposes is improbable.

Floors in industrial and retail buildings

Use

In the UK, industrial buildings are being increasingly constructed for use as storage and distribution facilities rather than for production. In these buildings the floor forms a vital part of the operation of the facility, and the specification of the floor must suit the tolerances of the storage and materials handling equipment which is to be installed.

Specification

The main factors which need to be considered in the specification of a floor are:

- ❖ durability and suitability of floor finish for intended use;
- ❖ tolerance on finished levels across the floor;
- ❖ accommodation of movement due to constructional and in-service movements; and
- ❖ type and intensity of loading.

In bespoke facilities, where the use of the floor and the location of plant or storage racking and aisles are known, the specification of all of the above can be quite precise. Where the final use of the building and the layout of the facility is not known, as is the case for speculative development, assumptions must be made, but it may still be beneficial in terms of costs and usability to define as much as possible before design of the floor commences.

Support

Structurally, floors in industrial buildings may be either ground supported or suspended. Suspended construction is where the floor is supported on piles and not designed to bear directly onto the ground immediately below the building. The design of such slabs is not dissimilar to the design of concrete upper floors in multi-storey buildings, with the concrete section being reinforced to span between supports.

The more common arrangement, ground conditions allowing, is the use of a ground-bearing slab. With this type of construction, the load capacity of the floor is derived largely from the strength of the underlying ground and the imported capping (sub-base). Reinforcement is included in the slab but does not contribute to strength, being essentially there to control cracking which, in an un-reinforced concrete section, would result from shrinkage as water in the freshly placed concrete dries out. The rate of drying and resulting shrinkage depends on many things including the method of curing, the thickness of the slab, the concrete mix design and the presence of a membrane under the slab.

Loadings

Floor slabs are subject to distributed loads, for example from materials stored on the floor, and to concentrated loads from materials handling plant, storage racking, mezzanine floors and internal division walls. Of these, it is generally the concentrated loads which are more critical to the design of the slab and hence there are moves to specify floor capacities in these terms rather than in terms of distributed loadings.

In general, the weakest parts of a floor slab are the edges and the sections immediately adjacent to joints, particularly where joints intersect.

Construction methods

The proposed construction method will influence the design of the slab. Increasingly, large floor slabs are constructed using specialised plant which only trade contractors specialising in floor construction, rather than general contractors, are likely to have. With this type of procurement and construction much of the design and specification of materials is done by the contractor. In particular, the use of fibre reinforcement instead of conventional reinforcement is common for large bay construction.

Design

Components

Concrete - the concrete mix will be specified by the structural engineer or specialist trade contractor to meet the requirements of the floor in terms of constructability, durability and finish. The concrete will include reinforcement in the form of either reinforcing bars or mesh, or fibres. Such fibres may be of steel or polypropylene.

Joints - joints are provided in a concrete floor slab to split it into easily constructed sections, and to cater for inevitable movement within the construction. This movement arises from shrinkage of the concrete soon after casting and from expansion and contraction due to changes in temperature over the life of the slab. Joints are expensive to provide and introduce weaknesses into the floor, and therefore, much design and specifying expertise has been expended into trying to reduce their number, or even to omit them entirely.

Membranes - the provision of a membrane immediately below the concrete slab can assist in reducing the friction which develops between the shrinking concrete as it dries out and the material below (sub-base). Good preparation of the sub-base and the inclusion of a membrane can reduce the amount of reinforcement required to control cracking. A membrane may also be used to minimise loss of fine particles from the concrete mix during placing and, if specified and laid appropriately, as a damp-proof membrane to protect against rising damp.

Sub-base - this is the imported material which forms the foundation of the floor slab and is necessary on all but the best natural ground (sub-grade).

Sub-grade - an investigation, undertaken on site with laboratory testing, to determine the nature of the ground on the site, is required to allow design of the floor slab. This investigation is normally undertaken by geotechnical firms to a specification that is usually drawn up by the structural engineer.

Flatness

Sophisticated materials handling equipment in warehouse and distribution facilities requires tight control of variation in level between different parts of the slab. In general terms, it is not economic to construct all concrete floor slabs to meet these requirements. Where tolerances are inappropriately specified, the slab may be unnecessarily expensive or unsuitable for certain uses.

Specialist surveying equipment is necessary to confirm compliance with the higher categories of level specification. Such surveys are usually undertaken soon after construction and records of survey results should be sought where flatness is important. Visual survey will not pick up failure of a floor slab to meet level tolerances unless the variation in level is extremely large.

Finish

In contrast to the majority of upper floors in buildings, the concrete often forms the finished surface of the floor in warehouse and retail buildings. Durability of the concrete surface against impact damage and wear, including tendency to form dust, is important.

Defects

Most defects in floor slabs affect the serviceability of the floor, but occasionally more fundamental structural defects occur. The most common defects are:

- ❖ dusting or breakdown of surface, or finishes;
- ❖ lack of flatness;

Design

- ❖ cracking;
- ❖ breakdown of joints; and
- ❖ structural defects due to slab or sub-base defects, or poorer than anticipated ground.

Increasingly, floor slabs will be considered as important components of buildings in the way that cladding and services are, and recorded information on aspects of their construction and performance will be required during the life of the building, particularly at times of change in use or ownership.

Further information may be found in 'Technical Report No 34 - Concrete industrial ground floors - A guide to design and construction' (ISBN 1 904482 01 5) published by the Concrete Society.

Building services design

- Air conditioning systems — 136
- Plant and equipment — 138
- Lift terminology — 139
- Lighting design — 140
- Data installations — 141

Design

Air conditioning systems

Air conditioning refers to a system or process for controlling within predetermined limits the temperature, humidity and sometimes the purity of the air in a building, accommodating the internal heat gains in conjunction with the external ambient conditions. The air, as well as being filtered, is heated or cooled as necessary and moisture is added or extracted to give a controlled humidity.

A comfort cooling system is essentially an air conditioning system but without full control of the humidity.

The table overleaf provides a comparison of some of the more common floor/ceiling system type characteristics. Wall positioned systems may be utilised however, these have different features. The abbreviations used in the table relate to the following system configurations:

DX:	Direct expansion split type unit system operating with a minimum fresh air system
VRV/VRF:	Variable refrigerant volume/flow system operating with a minimum fresh air system
Heat Pump:	Unitary reverse cycle heat pump system operating with a minimum fresh air system
FCU:	Four pipe fan coil unit system operating with a minimum fresh air system
VAV:	All-air variable air volume system incorporating terminal reheater batteries
Underfloor swirl:	All-air underfloor system with perimeter heating
Displacement:	Underfloor displacement system with ceiling positioned chilled beams and perimeter heating

Mixed mode ventilation systems

Mixed mode is a term used to describe engineering strategies that normally combine natural ventilation and/or mechanical ventilation and/or cooling in various combinations to achieve acceptable indoor environmental conditions in the most effective manner. It involves maximising the use of the building fabric and envelope to modify the internal climate/temperature swings, such as the use of night time pre-cooling of the building structure via the mechanical fresh air system. This approach has generally been used in offices; however, it is suitable for a wide range of building types.

Comparison of floor/ceiling based AC system type characteristics

	DX	VRV/VRF	Heat Pumps	FCU	VAV	Underfloor Swirl	Displacement
Installation costs	Low	Medium	Medium	Medium/high	High	Medium	Medium/high
Flexibility to change	Poor/average	Average/good	Poor	Good	Very good	Average	Average
Maintenance costs	Medium	Medium	Poor	Poor/medium	Good	Very good	Good
Operating costs	Medium	Low/medium	Low/medium	Poor	Very low	Low	Low
Ceiling void depth	Medium/low	Medium/low	Medium/high	Medium	High	Very low	Low
Floor void depth	Low	Low	Low	Low	Low	High	Medium
Suited to refurbishments	Average	Good	Average	Good/average	Poor	Poor	Poor
Resultant noise level	Average	Good	Poor	Average	Good	Good	Very Good
Plant space requirement	Medium	Medium	Medium	Low/medium	High	Medium	Low/medium
Riser space requirement	Low	Low	Low	Low/medium	High	Medium	Low/medium
Environmental control	Poor	Average	Poor	Good	Very Good	Good	Very Good

Design

Plant and equipment

Boilers – (purpose is to heat water for distribution)
- fuel can be gas, oil, coal, electric or dual fuel
- air is required for combustion and cooling (not for electric)
- some form of flue is required (not for electric)
- usually quiet in operation with insignificant vibration.

Water chillers – (purpose is to cool water for distribution)
- normally electrically driven, but can be gas driven
- conventionally air cooled to avoid the need for cooling towers that can be susceptible to bacteriological contamination (if near large body of water, water cooled chiller can be used)
- best externally located but can be internal and ducted to atmosphere
- refrigeration machines are complex and expert maintenance is required
- high noise and vibration levels can be generated
- old machines incorporated CFCs and HCFCs but new ones must not.

Air handling units – (purpose is to deliver heated/cooled and filtered air to various spaces)
- units incorporate fans, heaters, coolers, humidifiers and filters in various combinations
- may be located externally or internally with fresh air ducts to outside
- low tech equipment and easily maintained
- noise and vibration levels can be contained (but additional space required in plant room for silencers).

Diesel generators – (purpose is to provide a standby electrical supply to compensate for a breakdown in the main utility supply)
- normally diesel oil fed for commercial developments
- high fresh air and exhaust requirements demanding large louvred areas (for cooling)
- noise and vibration levels require special attention
- oil storage facility – daily use and possibly long term storage
- flue is required.

Design

Lift terminology

Power systems

Traction
- ideally requires motor room above (possible planning problem)
- can be positioned at other levels, usually below or adjacent to lift pit – more expensive (doubles load on structure)
- machine room - less (MRL) models now available
- incorporates counterbalance weight
- high efficiency
- high speed available.

Types
- single speed AC motor: up to 0.5m/sec jolt stop
- dual speed AC motor: up to 1.0m/sec, more accurate levelling
- geared variable voltage (VVAC and VVVF): smoother ride and greater speed (VVAC and VVVF)
- gearless variable voltages above 2.0m/sec, very quiet, long travel.

Hydraulic
- higher starting current, but not significant as compared to electric traction lifts
- maximum travel 20m
- less efficient, higher energy consumption
- no counterbalance
- limited starts per hour
- motor and pump house can be remote from shaft (up to 10m from lift pit). (Ventilation important for cooling)
- lower speed, up to 1.0m/sec.

Types
- direct acting: ram and bore hole below car
- side acting: Usually 'fork lift' action ram and cylinder within shaft
- indirect: combined ram and ropes, ram raises and lowers pulley.

Control systems

Automatic push button
- responds to first push – no calls stored; preference given to car button calls – flats and small offices.

Down collective
- answers landing calls in down direction only – mainly used for flats.

Full collective
- calls stored and answered in sequence in both up and down direction.

Full collective interconnected
- group of lifts interconnected as in full collective.

Programmed (or group collective) – tending to supersede other types.

Design

Door arrangement

Manually operated

Power operated

- ❖ single sliding: cheapest – larger shaft size required
- ❖ two speed side opening: expensive – minimum shaft size
- ❖ centre opening: quickest to full opening position.

Safety provisions

- ❖ retractable door safety edge – operating micro-switch
- ❖ pressure sensitive doors – operating micro switch on door operating mechanism
- ❖ electronic proximity detectors – detects presence of obstruction
- ❖ light ray or sonar.

For further information refer to BS EN81 Parts 1 and 2.

Lighting design

Categories of lighting, their respective lux levels, and the areas for which they are typically suitable are identified in the following table.

Category	Lux	Typical areas
Casual	100-150	Storage areas, plant rooms, lifts, circulation areas, bathrooms
Casual rough work	200-300	Dining areas, lounging rooms, bars, sports halls, libraries, rough machining
Routine work	300-500	General office, retail areas, lecture rooms, laboratories, kitchens, medium machining, supermarkets
Demanding work	750	Drawing offices, inspection of medium machining
Detailed work	1,000	Colour discrimination, fine machining and assembly, inspection rooms
Very fine work	1,500-3,000	Hand engraving, precision works, inspection of fine works

Lighting design should generally comply with the recommendations as laid down in the 2002 CIBSE Code for Lighting.

Lighting in areas where display screens are in use should comply with the requirements of CIBSE Lighting Guide LG3, including 'Addendum 2001'. The addendum covers new luminance limits for the performance of luminaires relating to various display screen types, in place of the old Cat 1, 2 and 3 standards which are now withdrawn. The purpose of the addendum is to promote consideration being given to the overall visual environment (ie surface reflections, direct daylight etc) and its effect on display screens and their users. Emphasis is also placed on designing schemes that avoid very high luminance patches in a space and abrupt changes in luminance across a surface or between adjacent surfaces.

A certificate of conformity to LG3 was introduced from May 2002 for completion by both the designers and installers. Copies of the certificate can be obtained from the CIBSE website.

Design

Data installations

Data installations and information technology require extensive cabling which, in turn, demands adequate access through the building. Therefore, it is important to consider access via risers, suspended floors or floor trunking in order to present the user with these services.

A raised floor is normally a basic pre-requisite to enable full flexibility of outlet positions. Minimum clear void typically 100 – 150mm depending upon floor plate size and riser arrangement.

Modern buildings are now usually block wired for voice and data using common telephone/data jack outlets.

Data/IT cabling is likely to require frequent upgrading and therefore components and component wiring needs to be accessible.

Site analysis

▶ Measured surveys 144

Design

Design

Measured surveys

It is strongly advisable to invest in measured surveys at all stages of a project, as early collection of the correct data will pay dividends.

The surveys are best carried out by a reputable measurement surveyor with independence and integrity. Most projects can be broken down into the following stages – the type of surveys recommended at each stage are listed.

Property acquisition – Check the floor areas independently (the RICS produce a booklet 'Code of Measurement' which is worth reading).

Check the lease or conveyance plans with the latest Ordnance Survey (OS) maps (1:1250 or 1:2500 scale). Check boundaries on site against legal documentation – resolve boundary disputes before completing the contract. The new Land Registration enables owners to determine their own boundaries and should be considered if any doubt arises.

Development – Existing measured surveys of the topography and/or buildings may exist. If they do, be sure to check their completeness and whether they are up to date. Consider employing a land surveyor to make checks. If none exist, then purchase the latest OS mapping (available digitally) for contextual purposes but be aware that the data will only be accurate to a metre or two.

It is strongly advised that you invest in a new measured survey. The RICS or TSA can provide assistance. The RICS has a document for large scale surveys which needs you to 'tick the boxes' to provide a specification. Also be aware that the more boxes you tick - the more cost is involved.

Various techniques are available now for obtaining data for adjoining properties such as large scale aerial photography and laser scanning (LIDDAR). This enables 3D data to be collected without gaining access to the property. Various 3D models of cities are now available on the internet but be very wary of their accuracy.

Architectural engineering and services, design and construction

The architect, engineer and services consultant will need measurements.

Everyone can and does take measurements. Modern equipment, such as Leica's 'Disto' and electronic theodolites make the task of surveying ever easier. It is strongly advisable though to seek the services of a good, reputable and honest land surveyor whose job it is to 'measure'. He will understand the complexities of the instrumentation and the necessary precision and accuracy for the task.

The culture within the property and construction industry is not to rely on others' dimensions and measurements. This all too often leads to disputes over measurements. It is worth considering employing an independent surveyor to collect the necessary measurements and supply these to all parties involved. A good surveyor will stand by his dimensions and this will lead to an improvement in time and less confrontation. Laser scanning technology provides an accurate 3D model instantly, from which measurements can be taken. This can be used to record progress and be displayed at design team meetings for all to see and take measurement from. Due to the fact this is 3D there is instant 'clash detection' (important for M&E Services), there is less ambiguity and misunderstanding than with 2D drawings and finally (and most importantly) it is understood by non-technical professions (eg, finance and legal advisers).

During construction it will be advisable to have a land surveyor on site, all the time if possible, or at least at regular intervals to ensure adequate dimensional control is established (in X, Y and Z planes) for the trade contractors to use.

Design

After completion a true 'as-built' survey is recommended. This can be used to check the area (gross and net) of the building and then be retained as part of the log book of the property. Ensure this is available in digital form to be used in CAD systems as well as a format suitable for Microsoft Office.

The plans will help the maintenance of the building and any future disposal. It is surprising how few reliable drawings exist.

An information system (like GIS) is worth considering to help the operation of the property. Each space can be allocated a unique reference number attached to the drawings to enable instant reporting. Additional data can be added about size, condition, occupancy, etc.

Computer-aided design

 Computer-aided design   148

Design

Computer-aided design

Computer-aided design (CAD) has become a widely used tool within the construction industry.

Both construction professionals and their clients now use CAD in the design, procurement and development of a variety of projects. It is this versatility, coupled with ever-reducing costs of hardware and software, that have increased the popularity of CAD and contributed to the decline of the traditional drawing board.

Hardware

The computing power required to run the most up-to-date CAD programs tends to vary between each package. The following system specification is for a typical basic installation, but advice should always be sought prior to purchase:

- ❖ Intel® Pentium® II or AMD K611 with a minimum processor speed of 450MHz.
- ❖ Microsoft® Windows® XP Professional, Microsoft® Windows® 2000 Professional, Windows 98, Windows ME or Windows NT® 4.0 (SP5 or later). (This software operates on a PC platform, however, many CAD packages have also been developed for use on a MAC platform).
- ❖ 128MB RAM and at least 200MB of free disk space (To install and run the software, further disk space will be required to save the drawing and backup files, either on the PC's hard drive or on a file server connected to a Wide or Local Area Network (WAN or LAN).
- ❖ VGA display 1024 x 768 or higher.
- ❖ CD ROM drive.
- ❖ Mouse or other pointing device (some CAD packages can be used with a tablet which provides the user with each command, symbol and block in view on the desktop for easy selection).

Higher specification PCs are available that will enable the user to operate the system more effectively and reduce the time spent to produce drawings.

Software

Since the initial development of desktop systems, a number of software manufacturers have established a market share in the supply of CAD programs, with the most commonly used being AutoCAD®, Microstation and ArchiCAD. Other programs exist which enable the user to produce enhanced drawings, such as 3D rendered visualisations and mechanical and electrical services drawings. Basic packages such as Volo™ View Express are available to download freely from the internet and enable PC users to view, mark up and print digital drawings received without having to purchase and understand a blown CAD program.

Further specialist software packages, which bolt onto the main CAD package, are available to support specialist uses. Examples include right to light calculation, daylight and sunlight analysis, sun path diagrams, digital measured and condition surveys, 3D development images, animations and fly-throughs.

Design

Principles of CAD

The production of two-dimensional drawings with the aid of CAD is based on the age old tradition of drafting on drawing board. The user still requires a good understanding of the principles behind producing a drawing, although the tools and processes used distinctly vary. Functions and terminology common to most CAD programs are listed below.

CAD files come in a variety of formats, such as DXF, DWG, DGN and DWF. When requesting digital drawings from an outside source, it is essential that a compatible format is specified. Drawings produced on a later version of the same software may not always be accessible through an earlier version.

Scale	Drawings are often prepared at 1:1 scale and then viewed on the title sheet within 'viewports' at a more appropriate scale, such as 1:50. The ability to produce a building or its components in this way increases the level of accuracy at which the drawing is produced and enables details to be generated from the 'master' drawing without having to redraw at an alternative scale.
Layers	Layers are used within CAD to enable different elements of a building's structure to be separated from another. For example, walls are placed on a different layer to windows. Layers can be turned on or off, removed from the printed copy, assigned a different colour, line thickness or style, without the user having to select each individual element or delete items from the drawing.
Styles	The styles in which a drawing is produced can be customised on a number of levels. They can be customised throughout a company, for an individual client, an individual project or down to the preferences of each user. This can include the style, the content and setup of each drawing, even the characterisation of the printed drawing copies, ensuring that each drawing issue maintains a uniformity in line with the degree of customisation implemented.
Templates	Templates enable the user to standardise drawing production. Many building product manufacturers now provide CAD-compatible 'blocks', which can be inserted into a drawing. Subscription services are also available, such as RibaCAD, which produce regular distributions incorporating the latest company templates and blocks.
Raster images	Raster images, can be incorporated into the drawing to present information which could not otherwise be drawn, for example, photographs.
Plotting	CAD drawings can be plotted at any scale and on any paper size. Standard paper sizes typically range from between A4 and A0, and scales range from 1:1 for intricate details through to 1:2500 for location plans. Drawings can be plotted in full colour, black and white and/or greyscales. Different line thicknesses can be used to highlight different elements of the drawing.

This list is not exhaustive and only provides a general overview of the simpler functions available in CAD.

The use of CAD in a project

CAD is usually implemented at project inception phase. This may be in the form of a digital survey of the development site and its surroundings. The project architect can then produce initial appraisal and feasibility studies prior to commencement of the full scheme.

Design

As the project develops, CAD can assist the entire project team in the production of project documentation. Compatibility of systems is required to ensure continuity across the team and to facilitate document exchange by e mail, project extranets and web-based file interchanges.

The ability to produce 3D images and models greatly improves presentation quality. Unlike a physical model, a 3D CAD model can be altered and manipulated with ease.

Throughout the construction phase, further working details can be produced and the construction drawings altered readily. Upon completion of the project, 'as-built' drawings can easily be produced and a full set of revisions can be stored.

Further information and hints

www.cadline.co.uk provides details of a large proportion of the CAD software currently available, including the sale and retail of software, training courses available and technical support for a number of CAD packages.

Tables and statistics

 Conversion formulae   152

Design

Conversion formulae

To convert to metric, multiply by the factor shown
To convert from metric divide by the factor shown

Length

miles: kilometres	1.6093
yards: metres	0.9144
feet: metres	0.3048
inches: millimetres	25.4

Area

square miles: square kilometres	2.59
square miles: hectares	258.999
acres: square metres	4046.86
acres: hectares	0.4047
square yards: square metres	0.8361
square feet: square metres	0.0929
square inches: square millimetres	645.16

Volume

cubic yards: cubic metres	0.7646
cubic feet: cubic metres	0.0283
cubic inches: cubic centimetres	16.3871

Capacity

gallons: litres	4.546
US gallons: litres	3.785
quarts: litres	1.137
pints: litres 0.568 gills: litres	0.142

Mass

tons: kilogrammes	1016.05
tons: tonnes	1.0160
hundredweights: kilogrammes	50.8023
quarters: kilogrammes	12.7006
stones: kilogrammes	6.3503
pounds: kilogrammes	0.4536
ounces: grammes	28.3495

To convert °C to °F

°C = 5/9 (°F) − 32
°F = 9/5 (°C) + 32

Materials

- Deleterious materials — 154
- Asbestos — 156
- High Alumina Cement concrete — 161
- Chlorides — 163
- Corrosion of metals — 165

Materials and defects

Deleterious materials

The presence of deleterious materials in a building may affect its market value and could, in severe cases, result in element failure or affect the health of persons working or living there.

The reaction of investing institutions to these materials depends on a number of factors and often the presence of a deleterious substance will not prevent a purchase. However, great care must be taken to assess the actual risks or consequences involved, so that a value judgement can be made.

Materials hazardous to health

The more common, hazardous materials, and associated risks, are identified in the following table.

Materials	Common use	Use risk
Lead	When used in water pipes and lead paint (lead roofing materials pose little or no risk).	Risk of contamination of drinking water in lead pipes, or from lead solder used in plumbing joints. Risk of inhalation of lead dust during maintenance of lead based paint. Risk to children of chewing lead painted surfaces (Pica). Concentration of lead in paint now generally much reduced. Beware of lead content in brass fittings.
Urea Formaldehyde foam	Cavity wall insulation. Some insulation boards but rare in UK.	There is some evidence that UF foam may be a carcinogenic material although this is not proven. Vapour can cause irritation. Poorly installed insulation can lead to passage of water from outer leaf of brick to inner leaf in cavity wall situation. There are some worries over formaldehyde used as an adhesive in medium density fibreboard and chipboard but this is likely to be a problem only in unventilated areas with large amounts of boarding.
Asbestos See also page 156	Commercial and residential buildings as boarding, sheet cladding, insulation and other uses particularly in the 1950s, 60s and 70s.	Airborne asbestos fibres may be inhaled and eventually lead to either asbestosis, lung cancer or mesothemelioma.

Materials and defects

Materials damaging to buildings

Those materials which may affect building performance or structure are identified in the following table.

Materials	Common use	Use risk
Calcium silicate brickwork	Used in lieu of concrete or clay bricks, often as an inner leaf in cavity work. Often cited as deleterious but if used correctly will perform well.	Brickwork shrinks after construction with further movement due to wetting. Construction must provide measures of control to distribute cracking. Concrete bricks may display a similar propensity to shrinkage and again care must be taken in the design of movement joints etc.
Calcium chloride concrete additive	Commonly used in in-situ concrete as an accelerator and often added in flake form. Often found in buildings constructed before 1977. (May also be present from atmospheric or traffic exposure).	Reduces passivity of concrete in damp conditions. Subsequent risk of corrosion of steel reinforcement.
High Alumina Cement (HAC) Further details on page 161	Mainly used in the manufacture of pre-cast X or I roof or floor beams together with some lintels, sill members etc between 1954 and 1974. HAC has existed since about 1925.	Strength of concrete can decrease significantly often when high temperatures and/or high humidity is involved. Defects may be due to faulty manufacture.
Sea dredged aggregate	In-situ concrete or pre-cast concrete.	May contain salts such as sodium chloride. If salts not properly washed out, risk of corrosion reinforcement sodium may contribute to alkali silica reaction. Provided the aggregates are properly washed and controlled in accordance with British Standard requirements the indications are that there are no greater risks involved than with the use of aggregates from inland sources.

Materials and defects

Materials	Common use	Use risk
Mundic blocks and Mundic concrete	Concrete blocks and concrete manufactured from quarry shale commonly found in the West Country.	Loss of integrity in damp conditions. Further research required to identify level of risk across the country.
Woodwool slabs (also woodcrete and chipcrete)	Often used as (a) decking to flat roofs, or (b) as permanent shuttering.	Use in (a) may be considered acceptable providing material is kept dry. Use in (b) as a permanent shutter may result in grout loss (honeycombing) or voidage of concrete near to or surrounding reinforcement, particularly with ribbed floors. May result in reduced fire resistance, reinforcement corrosion or in extreme cases loss of structural strength. May be repaired by application of sprayed concrete. Condition investigated by cut-out removal of woodwool in many locations.
Brick slips	Typically 1970s and 1980s to conceal flow nibs in cavity walls.	Risk of poor adhesion, lack of soft joints can transfer load to slips and cause delamination.

Asbestos

Asbestos is the generic term for several mineral silicates occurring naturally in fibrous form. Because of its various useful properties (resistance to heat, acids and alkalis, and good thermal, electrical or acoustic insulator) it has been extensively used in the construction industry.

Three main types in the UK are Chrysotile (white), Amosite (brown) and Crocidolite (blue). It is used in a variety of forms varying from boards or corrugated sheets to loose coatings or laggings and, generally, the more friable the material, the greater the asbestos content.

Health risk

Inhalation of its microscopic fibres can constitute a serious health risk and is associated with several terminal diseases.

Not all asbestos, irrespective of its circumstances, constitutes an immediate risk, although the effects of possible future disturbance or deterioration must be considered. Factors to be taken into account are the type, form, friability, condition and location of the source material.

The (1987) Joint Central and Local Government Working Party on Asbestos concluded that "Asbestos materials which are in good condition and not releasing dust should not be disturbed... Materials that are damaged, deteriorating, releasing dust or which are likely to do so should be sealed, enclosed or removed as appropriate. Materials which are left in place should be managed and their condition periodically reassessed. The risk to the health of the public from asbestos materials which are in sound condition and which are undisturbed is very low indeed. Substitute

Materials and defects

materials should be used where possible, provided they perform adequately".

The HSE actively discourages the unnecessary removal of sound asbestos materials and each case should be decided on its own merits following an assessment of the risks arising.

In the past the people most at risk have been workers in the asbestos industry involved in the importation, storage, manufacture and installation of materials or components containing this material.

These activities are now banned in the UK and the removal, treatment or intentional working with asbestos is strictly controlled and generally limited to specialists. The risk however continues for anyone who inadvertently disturbs the asbestos in the course of their routine business, including builders, maintenance workers, electricians and the like, and the most recent legislation is intended for their protection.

Legislation

The enabling act for asbestos legislation is the 1974 Health & Safety at Work Act and failure to comply with the requirements is a criminal offence.

There are two principal regulations that apply to works that could expose persons to the risk of respirable asbestos fibres.

The Asbestos (Licensing) Regulations 1983 (as amended)

Broadly, these regulations require any person who manages or carries out work with asbestos insulation, asbestos coating or asbestos insulating board, or the clearance of asbestos contaminated land, to hold a licence that is issued by the HSE.

The Control of Asbestos at Work Regulations (CAWR) 2002

These consolidated similar previous asbestos regulations and they also introduced a new duty to 'manage asbestos' (Regulation 4) together with a requirement to use only accredited persons to analyse the content of materials suspected of containing asbestos (Regulation 20).

With the exception of regulations 4 and 20, CAWR became effective from 21 November 2002. Regulation 4 came into force on 21 May 2004 and regulation 20 on 21 November 2004.

Notwithstanding the requirement for a licence, CAWR applies to any work which **may** involve the exposure of persons to **any** form of asbestos irrespective of the type, form or amount of asbestos and includes activities involved in both sampling and laboratory analysis.

Broadly, and using the number of the appropriate regulation as a prefix, they require:-

a) The 'Dutyholder' to:
4   manage asbestos in non-domestic premises.

b) 'Every person' to:
4 (2)   cooperate with the dutyholder so far as is necessary to enable him/her to comply with their duties to manage asbestos in non-domestic premises.

c) The employer to:
5   identify the type of asbestos;
6   prior to the works, assess the likely level of risk, determine the nature and degree of exposure and set out steps to prevent or control it;
7   produce a suitable written plan of work;
8   *notify the Enforcing Authority of non-licensable work (at least 14 days before works);
9   inform, instruct and train personnel;

Materials and defects

10. prevent exposure of employees to asbestos so far as is reasonably practicable and where not reasonably practicable, reduce to lowest level reasonably practicable, both the exposure (without relying on the use of respirators) and the number of employees exposed;
11. ensure that any control measures are properly used or applied;
12. maintain control measures and equipment (keeping records of the latter);
13. provide suitable personal protective clothing and ensure that it is properly used and maintained;
14. establish arrangements to deal with accidents, incidents and emergencies;
15. prevent spread of asbestos from the workplace, or where not reasonably practicable, reduce to lowest level reasonably practicable;
16. keep asbestos working areas and plant clean and thoroughly clean on completion;
17. *designate 'asbestos areas' and 'respirator zones'. Monitor and record exposure where appropriate;
18. *monitor exposure of employees to asbestos by air monitoring (if not deemed necessary record reasoning);
19. ensure that air testing is only carried out by a person with ISO 17025 accreditation;
20. ensure that sampling of material to determine whether it contains asbestos is only carried out by a person with ISO 17025 accreditation.
21. *maintain health records and medical surveillance of employees.
22. Provide suitable washing and changing facilities (in addition to general welfare); and
23. ensure all raw materials or asbestos waste is stored, received into, dispatched from or distributed within suitable sealed, clearly labelled containers.

   * These specific requirements are only triggered if the likely level of exposure to respirable fibres exceeds stated limits.

The Regulations establish these trigger points in the form of 'action levels' and 'control limits'.

Action levels

If the action level is likely to be exceeded, the following will apply:-

- ❖ Written notification of Enforcing Authority (14 days prior to commencing work).
- ❖ Designation of 'asbestos working areas' where access is limited to authorised personnel.
- ❖ Regular medical surveillance of employees with health records kept for at least 40 years from the date of the last entry.

The action levels refer to accumulative exposures, as identified in the following table, whereby the respirable asbestos fibres per millilitre of air are multiplied by the number of hours of exposure over a continuous 12-week period.

Type of asbestos	Fibre-hours per ml
White	72
Any other type or mixture	48

Control limits

If the control limit is likely to be exceeded, the following will apply:-

- ❖ Designation of 'respirator zones' where suitable respirators must be worn at all times.
- ❖ Suitable respirators and protective equipment must be issued, used and maintained.

Materials and defects

The control limits are identified in the following table.

Type of asbestos	Average/measured in fibres per ml over a continuous period	
	four hours	ten minutes
White	0.3	0.9
Any other type or mixture	0.2	0.6

"Duty to manage asbestos in non-domestic premises" (regulation 4)

The 'dutyholder' responsible for the management of asbestos in non-domestic premises as set out in regulation 4(1) is every person, who has by virtue of a contract or tenancy, an obligation for its repair or maintenance, or, in the absence of such, control of those premises or access thereto or egress therefrom.

This includes those persons with any responsibility for the maintenance, or control, of the whole or part of the premises.

When there is more than one dutyholder, the relative contribution required from each party in order to comply with the statutory duty, will be shared according to the nature and extent of the repair obligation owed by each.

This regulation does not apply to 'domestic premises', namely a private dwelling in which a person lives, but legal precedents have established that common parts of flats (in housing developments, blocks flats and some conversions) are not part of a private dwelling.

The common parts, are classified as 'non domestic' and therefore regulation 4 applies to them, but not to the individual flats or houses in which they are provided.

Typical examples of common parts are entrance foyers, corridors, lifts, their enclosures and lobbies, staircases, common toilets, boiler rooms, roof spaces, plant rooms, communal services, risers, ducts and external outhouses, canopies, gardens and yards.

The regulation does not however apply to kitchens, bathrooms or other rooms within a private residence, that are shared by more than one household, or communal rooms within sheltered accommodation.

The duties are identified in the following table.

Subject	Requirement
*Cooperate	Cooperate with other dutyholders so far as is necessary to enable them to comply with their Regulation 4 duties.
Find and assess condition of ACMs	Ensure that a suitable and sufficient assessment is made as to whether asbestos is or is liable to be present in the premises and its condition, taking full account of building plans or other relevant information, the age of the building and inspecting those parts of the premises which are reasonably accessible. (Must presume that materials contain asbestos unless strong evidence to the contrary). (see MDHS 100 for guidance on asbestos surveys).

Materials and defects

Subject	Requirement
Review	Review assessment if significant change to premises or suspect that it is no longer valid and record conclusions of each review.
Records	Keep an up-to-date written record of the location, type (where known), form and condition of ACM's.
Risk assessment	Where asbestos is or is liable to be present assess the risk of exposure from known and presumed ACM's.
**Management plan	Prepare and implement a written plan, identifying those parts of the premises concerned, specifying measures for managing the risk including adequate measures for properly maintaining asbestos or where necessary, its safe removal.
Provide information to others	Ensure the plan includes adequate measures to ensure that information about the location and condition of any asbestos is provided to every person likely to disturb it and is made available to the emergency services.
Review and monitor	Regularly review and monitor the plan to ensure it is valid and that the measures specified are implemented and that these are recorded.

** Management plan

(see HSE publication 'A comprehensive guide to managing asbestos in premises' HSG227).

The management plan is an important legal document which in addition to its health and safety significance will be required to be made available to, and inspected by, a variety of interested parties.

The absence of such a document may thus have significant financial implications or affect the liquidity of the premises as an asset.

A plan is not required when the assessment whether asbestos is or is liable to be present in the premises confirms that it is not. For example, the building is new and there is confirmation from the project team that asbestos has not been used in its construction.

Nevertheless a record must be kept of the assessment carried out and its conclusion to show to an inspector or prospective purchaser or occupant.

The dutyholder owns and is responsible for the safekeeping of the plan, however he is obliged to make the information available "at a justifiable and reasonable cost" to anyone who is likely to disturb asbestos and this includes new owners or occupants.

Duty to cooperate

'Every person' has a duty to cooperate with the dutyholder so far as is necessary to enable the dutyholder to comply with his duties under Regulation 4.

This includes the landlord, tenants, occupants, managing agent, contractors, designers and planning supervisor.

Materials and defects

The possible scenarios envisaged by the ACOP include:-

- ❖ Anyone with relevant information on the presence (or absence) of asbestos.
- ❖ Anyone who controls parts of the premises to which access will be necessary to facilitate the survey and management of asbestos (ie its removal or treatment or periodic inspection).

Cooperation does not extend to paying the whole or even part of the costs associated with the management of the risks of asbestos by the dutyholder(s), who must meet these personally.

Where there is more than one duty holder for a premise the costs of compliance will be apportioned according to the terms of any lease or contract determining the obligation to any extent for its repair and maintenance.

If there is no such documentation then the apportionment of costs will be based on the extent to which parties exercise physical control over the premises.

In the final analysis, the courts will decide financial responsibility using the principles outlined above.

Guidance in the ACOP states that architects, surveyors or building contractors who were involved in the construction or maintenance of the building and who may have information that is relevant "would be expected to make this available at a justifiable and reasonable cost".

The duty to cooperate is not subject to any limitation or exclusion, thus there is an obligation to do whatever is necessary to cooperate with the dutyholder.

Short lease tenants, licensees or other occupants who control access, but do not have any contractual maintenance liabilities would be required to permit the landlord access to fulfil his/her duties.

In June 2003, RICS published a guidance note 'Asbestos and its implications for members and their clients' (ISBN 1-84219-063-6).

High Alumina Cement concrete

Background

The manufacture of High Alumina Cement (HAC) commenced in the United Kingdom in 1925 to provide concrete that would resist chemical attack, particularly for marine applications. This cement developed high early strength, although its relatively high cost prevented extensive use.

During the late 1950s and 1960s the main use of HAC was for the manufacture of precast prestressed components which could be manufactured quickly, therefore offsetting the additional cost of the material.

The earliest UK failures were experienced during 1973-74 when school roofs collapsed. There followed considerable investigation and testing. In 1975 the Building Regulations Advisory Committee [BRAC] Sub-Committee P published design criteria to be used in checking the adequacy of buildings containing HAC structural members. The two main aspects of the investigation were, firstly, a strength assessment and, secondly, a durability assessment.

In 1976 HAC concrete was banned for structural use.

Problems with HAC

HAC concrete undergoes a mineralogical change known as conversion. This conversion is accompanied by a loss of strength and increased porosity. Consequently there is also a reduction in resistance to chemical

Materials and defects

attack. The higher the temperature during the casting of the concrete, the more quickly conversion takes place.

The relationship between conversion and strength is complex, however the strength of highly converted concrete is very variable and is substantially less than its initial strength. Typically the original design strength of 60N/mm2 may be reduced to 21N/mm². This lower figure represents a residual strength below which no further loss of strength will occur during the remaining life of the material.

Highly converted HAC concrete is vulnerable to acid, alkaline and sulphate attack. For this to take place water as well as the chemicals must have been present persistently over a long period of time at normal temperatures. Chemical attack is usually very localised in nature and the concrete typically degenerates to a chocolate brown colour and becomes very friable, often due to sulphate attack.

Given the sensitivity to moisture the greatest risk therefore lies in the use of roof members. It is therefore important to appraise the condition of the concrete and any waterproof coverings before making any formal judgement as to the remedial work required.

In a warm and moist environment there is the possibility of chemical action occurring where high alkali levels may be present as a consequence of the use of certain types of aggregate or where alkalis may have ingressed from plasters, screeds and woodwool slabs. In such circumstances, HAC is vulnerable to chemical attack.

Investigation

There are three generic stages in an investigation, namely:-

- ❖ Stage 1 - identification
- ❖ Stage 2 - strength assessment
- ❖ Stage 3 - durability assessment.

A Stage 2 strength assessment is required to determine if the precast concrete members have sufficient structural capacity, even at the reduced fully converted strength, to safely withstand the applied loading. Sub-Committee P sets out the guidelines for concrete strength based on 21N/mm2. The strength assessment requires the section properties of the beam to be established. Thereafter the structural strength of the element can be calculated. In cases where the section properties are unknown, or cannot be determined by investigation, then assessments are limited to determining the concrete strength using near-to-surface tests.

A Stage 3 durability assessment is required to determine the long-term durability where affected by chemical attack and reinforcement corrosion. Testing can be undertaken to determine the presence of alkalis and sulphates. Laboratory testing may be supplemented by a detailed visual inspection and the removal of lump samples for petrographic examination. A durability assessment should also include a visual examination of the reinforcing steel where lump samples are removed. In recent years it has been found that HAC is less durable then members containing ordinary Portland cement.

Assessment

Putting all of this in context, however, there have been no recorded instances in this country of a failure of a floor incorporating HAC concrete. It is important to know that in the case of the original historic failures, manufacturing faults were eventually discovered. The greatest reduction in strength occurred where a high water content was present during the period of mixing and high temperatures took place during curing.

Of the five failures or new failures of roof constructions, two did not directly involve the quality of the concrete, one was aggravated by chemical attack and two were apparently due to defective concrete which should have been rejected at the time of casting. On no occasion has weakening of the concrete due to conversion been the sole cause of failure.

Chlorides

The presence of chlorides, whether added as calcium chloride or ingressed as de-icing salts, may result in 'chloride induced corrosion' and is less common than corrosion caused by low cover. When chloride corrosion does occur its effects may be wide ranging including a reduction in structural capacity.

Natural alkalinity of concrete

Steel does not corrode when embedded in highly alkaline concrete despite high moisture levels in the concrete because a passive film forms on the steel and remains intact as long as the concrete surrounding the bar remains highly alkaline.

Chloride induced corrosion

Corrosion may occur in concrete that contains sufficient chlorides even if it is not carbonated or showing visible signs of deterioration.

The presence of free chloride ions within the pore structure of the concrete interferes with the passive protective film formed naturally on reinforcing steel.

Chloride ions exist in two forms in concrete namely free chloride ions, mainly found in the capillary pore water, and combined chloride ions which result from the reaction between chloride and the cement hydration process. These occur in proportions that depend on when the chloride entered the concrete. If chloride was introduced at mixing, for example, as calcium chloride, approximately 90% may form harmless complexes leaving only 10% as free chloride ions. If, on the other hand, sea water or de-icing salts penetrate the surface of the concrete, the ratio free to combined chloride may be 50:50.

The corrosive effect of chlorides is significantly affected therefore by the presence of free chlorides. The effects of chlorides are classified in terms of risk of corrosion because in certain conditions even low levels of chloride may pose some risk. The permissible level of chloride added at mixing specified in BS 8110 is 0.4% by weight of cement. For pre-stressed concrete the level is lower at 0.06%. The overall effect of reinforcement corrosion caused by chlorides must therefore be considered with the depth of reinforcement and the depth of carbonation.

There are two methods by which chlorides can be the cause of corrosion in reinforced concrete:-

- ❖ As cast-in chloride, usually calcium chloride added at the time of mixing or from sea-dredged aggregates.
- ❖ As ingressed chloride, ie by the penetration of the outer surface of the concrete from de-icing salts.

Chloride induced corrosion results in localised breakdown of the passive film rather than the widespread deterioration that occurs with carbonation. The result is rapid corrosion of the metal at the anode leading to the formation of a 'pit' in the bar surface and significant loss of cross sectional area. This is known as 'pitting corrosion'. Occasionally a bar may be completely eaten through.

Chloride induced reinforcement may occur even in apparently benign conditions where the concrete quality appears to be satisfactory. Even if

Materials and defects

there is poor oxygen supply reinforcement corrosion may still take place. Failure of reinforcement may therefore occur without any visual sign of cracking or spalling.

All aggregates used commonly for concrete mixing contain a background level of chlorides usually less than 0.06% by weight of cement. The use of calcium chloride as an accelerating additive at the time of mixing was popular during the 1950s and 1960s. It was used in precasting yards to speed up the re-use of expensive moulds and was used on site during cold weather to increase the rate of gain in strength. The use of calcium chloride was banned in 1977. The presence of calcium chloride cast-in within the mix usually attracts a chloride level significantly greater than 0.4% by weight of cement. Ingressed chlorides through the outer surface of the concrete are variable in nature. However the use of de-icing salts on, for example, external staircases and balconies is popular and may result in localised high concentrations of chlorides in excess of 1.0% by weight of cement. Concentrations of ingressed chlorides on the top surface of a car park deck may typically occur up to 3 or 4%.

The presence of calcium chloride is further exaggerated by the presence of deep carbonation. Carbonation releases combined chlorides into solution to form free chloride ions, thus increasing the likelihood of corrosion. For this reason many properties built approximately 20-30 years ago may only now start causing problems.

Risk assessment

Guidance on assessing the risk of reinforcement corrosion is provided by the BRE in Digest 444 Part 2. Here the risk of corrosion for structures of various ages is presented in the range negligible, to extremely high risk. Factors effecting the risk assessment are either a dry or damp environment, the depth of carbonation and of course the level of chlorides present.

Repair

The successful repair of chloride induced corrosion is notoriously difficult because of the tendency for new corrosion cells to form at the boundary of the repair. This mechanism is called 'incipient anode effect' and should be minimised by removing, wherever possible, all concrete with significant chloride contamination. In recent years the introduction of proprietary sacrificial zinc anodes embedded within the patch repair and attached to the reinforcement can help to reduce this effect. For high levels of chlorides and long-term protection this may not be sufficient.

For heavily chloride contaminated structures, particularly car parks, the only tried and tested long term solution is cathodic protection. The cost and complexity of installing cathodic protection is not usually warranted within building structures. A variation of cathodic protection is desalination – a short-term process using higher current densities than cathodic protection. Migrating corrosion inhibitors have found some success; these are penetrating surface coatings applied under strict conditions.

Materials and defects

Corrosion of metals

Bi-metallic corrosion

Bi-metallic or galvanic corrosion is experienced when two dissimilar materials are in electrical contact and are bridged by an electrolyte. The electrolyte could be water containing salt, acid or a combustion product. The electrolytic cells comprise a series of positive anodes and negative cathodes between which current flows and at which electrochemical reactions take place. The degree to which a metal is subject to this form of attack is determined by the difference in voltage potential between two metals, the amount of moisture present, the relative areas of contact, the corrodent concerned and whether either or both of the metals have naturally occurring oxide films.

If it is necessary to use dissimilar materials they should be isolated with washers of a non-conductive material, for example, neoprene, PTFE, SRBF etc. Painting of the contact surfaces with bitumen is an alternative, but less reliable solution.

The electro-chemical series

The further apart two metals appear in the following table the more actively the two metals will react when placed in contact in a slightly acid aqueous solution.

Metal	Chemical symbol	Normal electrode potential (volts)
"Noble" or Cathodic, ie, protected end		
Gold	Au	+1.42
Platinum	Pt	+1.20
Silver	Ag	+0.80
Mercury	Hg	+0.80
Copper	Cu	+0.345
Lead	Pb	-0.125
Tin	Sn	-0.135
Nickel	Ni	-0.24
Cadmium	Cd	-0.40
Iron	Fe	-0.44
Chromium	Cr	-0.71
Zinc	Zn	-0.76
Aluminium	Al	-1.66
Magnesium	Mg	-2.38
Sodium	Na	-2.71
Potassium	K	-2.92
Lithium	Li	-3.02
"Base" or Anodic, ie corroded end		

Effects of bi-metallic combinations

For cladding supports, aluminium alloys are often employed. When aluminium and stainless steel are in contact, there is the potential for corrosion to occur. The extent of corrosion will depend upon the respective sizes of the two metal components.

For example large aluminium fitting/small stainless steel bolt: little corrosion.

Building defects

Materials and defects

- Common defects in commercial and residential properties — 168
- Problem areas with 1960s and 1970s buildings — 170
- Defects in concrete — 184
- Fungi and timber infestation in the UK — 185
- Troublesome plant growth: Japanese Knotweed — 189
- Rising damp — 191
- Rising groundwater — 193
- Subsidence — 194

Materials and defects

Common defects in commercial and residential properties

Period	Typical problem	Possible effects
Pre 1900	Failed or lack of damp proof course.	Rising dampness, penetrating damp, efflorescence on plaster, decay to skirtings and the like.
	Poor ventilation of floor voids.	Decay in wall plates, joists etc.
	Poorly fitting sash windows, risk of decay within window reveals, water penetration beneath sub sills.	Draughty or dangerous operation, decay in concealed areas, lack of security.
	Damp penetration through 225mm brick walls.	Damage to plaster and finishes, decay to wall plates and bonding timbers.
	Poor quality repairs to roofs and gutters.	Risk of timber decay.
	Roof covered with concrete interlocking tiles.	Overloading of roof structure, bowing of rafters and purlins, roof spread.
	Lack of restraint to flank walls.	Bulging or instability, associated cracking on front and rear elevations.
	Settlement of internal partitions.	Plaster damage, distortion in floors and door openings.
	Alterations to trussed (loadbearing) partitions.	Removal of support gives rise to distortions in floors, reduction of loadbearing capacity and possible risk of collapse.
	Failure of brick arches and timber lintels.	Cracking and distortion of brickwork above window heads.
	Defective rainwater goods.	Risk of decay in built-in timbers.
	Settlement of bay windows.	Internal cosmetic damage, distortion in loadbearing elements.
	Insect attack, particularly in poorly ventilated and damp areas such as floor and roof voids.	Loss of strength if particularly badly affected.
	Lead water mains.	Possibly hazardous to health - partly depends on plumbsolvency of the water.
	Delamination of brick skins.	Bulging of brickwork.

Materials and defects

Period	Typical problem	Possible effects
1900-1939	Wall tie failure in cavity brickwork.	Bulging of brickwork, horizontal cracking or 'pagoda effect'.
	Delamination of render finishes to walls.	Cracking and bulging of render, detachment of same.
	Corrosion of roofing nails.	Slipping of tiles.
	Lead water mains.	Possibly hazardous to health.
	Outdated electrical services.	Possibly dangerous.
	Corroded rainwater goods.	Risk of decay in built in timbers, damp penetration.
	Corroded galvanised steel or steel windows.	Cracked glazing, high maintenance costs.
	Timber joinery.	Decay to calls and softwood frames.

Period	Typical problem	Possible effects
All	Over notching of floor joists.	Reduction in strength, sagging.
	Removal of chimney breasts.	Possible lack of support.
	Provision of insulation, blocking ventilation paths.	Condensation.
	Blocking of airbricks.	Lack of ventilation, risk of decay.
	Removal of loadbearing or walls affording stability.	Possible long term structural consequences.
	Removal of, or planting of trees or large shrubs.	Possible desiccation or re-hydration of sub soil, damage to drains or foundations.
	Replacement windows.	Poor support to bay windows, distortion of brickwork.

Materials and defects

Problem areas with 1960s and 1970s buildings

The 1960s saw a frantic period of innovation and experimentation. The need for replacement housing following the Second World War gave rise to the development of numerous system types using innovative materials. Construction had started to drift toward an assembly process and away from traditional craft-based skills.

Some 1960s construction was appallingly bad, for example, some high rise social housing, but other schemes were of a very high standard using good quality and durable materials. Many buildings are now facing or have undergone major refurbishment or change of use (for example, office to housing or hostel accommodation), but many examples of good and bad construction remain.

Rather than look at typical problems associated with each style of building, this list overleaf is intended to give a very brief introduction to a number of common or typical materials or construction faults. The list is not exhaustive.

Brief resume of some problems

Key to common building types - these are indicative only

W = warehouse or industrial

O = Office or commercial developments

H = Housing

A = All types.

Materials and defects

	Item, element or material	Effects
1	**Aluminum sash windows (O)** A common type of window was the vertical sliding sash. Instead of the vision glazing being held in a frame, the glass ran in aluminium tracks, with horizontal top and bottom frame members clipped onto the glass. Spring balances were used to hold the windows open.	By now, these windows will be very worn. Defects in the springs or breakage of the glass can lead to the ejection of an entire sash window - clearly a health and safety issue. Treat these windows with caution.
2	**Asbestos (A)** Very common in 1960s buildings. Chrysotile (white) for some insulation boards, roof sheets, water tanks, cill boards etc. Artex, floor tiles, partition wall linings, fire doors etc may also have a content. Amosite (brown) used as insulating boards, fire protection or fire breaks, behind perimeter heaters, partitions etc. Crocodilite (blue) often paste-applied friable material in boiler rooms, pipework, calorifiers, etc.	Major health risks depending on type, location risk of disturbance. Deleterious material, detection, management and control are highly regulated. Ask for a copy of the asbestos register.
3	**Asphalt roofs (O)** These could be of quite good quality and may have performed well if laid on a concrete deck.	The lack of insulation would have helped to reduce temperature ranges and so restrict thermal movements.
4	**Calcium chloride concrete additive (A)** Often used by manufacturers of pre-cast elements or for concreting in cold weather. Enables rapid set and removal of moulds. Can also be found in brickwork mortar. Use of un-washed sea dredged aggregates may have led to chloride contamination.	Considered a deleterious material. Creates conditions of high electrical conductivity in the concrete with consequent high risk of corrosion of steel reinforcement. Very difficult to repair effectively. Look out for spalling concrete and very black, stained steel. Can result in severe pitting corrosion without disruption of the surface of the concrete. Tests should always be recommended. Since about 1978, this material should not have been used. In brickwork, risk of corrosion of wall ties. Exposure to de-icing salts, salt spray etc can also be very damaging. Watch out for coastal locations, car parks, road bridges etc.

Materials and defects

	Item, element or material	Effects
5	**Calcium silicate bricks (A)** A smooth, often creamy coloured brick made from lime, sand and flint. Small particles of flint can sometimes be seen in cut bricks or weathered surfaces. Can be mistaken for concrete bricks (see below). Widespread use in 1960s and 1970s and still manufactured and gaining popularity again.  Flank wall of calcium silicate bricks - shrinkage cracking.	Prone to shrinkage (unlike clay bricks which expand after laying). If movement control joints are missed or badly spaced (which they often were) diagonal cracking can occur. Thermal or moisture cracking often visible at changes in the size of panels, for example long runs below windows coinciding with short sections between windows. Look out for thin bed cracks and wider cracks to vertical joints. Do not confuse with subsidence cracking or corrosion of steel sub frame. Use as a backing to clay brickwork likely to cause problems as a result of expansion of clay brick and contraction of calcium silicate brick.
6	**Cold bridging and condensation (A)** Poor insulation standards led to problems with severe cold bridging particularly in housing where humidity levels are higher. Polystyrene insulation was sometimes used but this was usually no more than 25mm thick. In the mid to late 1970s thermal insulation standards were increased, particularly in industrial buildings.	Watch out for cold bridging around balcony structures and pre cast lintels. Provision of insulation within industrial buildings resulted in a spate of condensation problems within roofs. Cold night sky radiation gave rise to condensation on the underside of metal roofing, while poor application of vapour control layers meant that problems were exacerbated.
7	**Cold flat roof construction (A)** Little thought was given to vapour control or for that matter roof insulation. It was common to provide sealed flat roof construction with minimal insulation and sometimes a foil backed plasterboard ceiling lining. Ventilation to the roof void was often ignored. Built up felt roofs were often asbestos based, but had a life of around 15 years and no more. For this reason most felt roofs would have been replaced by now.	Risk of condensation occurring, with subsequent risk of decay to roof decking or to structure. 

Materials and defects

	Item, element or material	Effects
8	**Concrete (A)** Can be of mixed quality, sometimes poorly compacted and with lack of cover to steel reinforcement. Under codes, depth of cover for external work should have been circa 40mm.	Sometimes poor durability, corrosion due to the effects of carbonation or chloride content. Tests should be recommended. Poor curing methods could mean lack of durability. Calcium chloride added as an accelerator either in pre-cast or insitu work.

Waffle slab showing the effects of chloride attack.

9	**Concrete boot lintels (A)** Concrete lintels designed to have a projecting nib to support the outer leaf, and built only into the inner leaf, to provide a neat appearance externally.	Rotation of the lintel under eccentric load, creating diagonal cracking to the brickwork above the window. Other signs are opening of the bed joint immediately above lintels and splitting of the reveal brickwork immediately beneath. Once rotation has taken place, brickwork will tend to arch over the opening, thus relieving some of the load on the lintel. Cracks can then be repointed.
10	**Concrete bricks (A)** Similar in appearance to calcium silicate brick but often used in dark brown, dark red or dark grey variants. Harder and coarser texture than cs bricks.	Suffer from similar shrinkage related problems. Can be hard to differentiate between these and calcium silicate bricks, but may be harder and contain small particles of visible aggregate.
11	**Corrugated 'big six' asbestos cement sheet (W)** Often found on industrial buildings and warehouses well into the 1970s. Name given as a result of the 6-inch profile, but in fact 'big six' was one of several different profiles of sheet. Often based on an asbestos content of around 12-15% chrysotile, (white asbestos) with profiled eaves and ridge pieces and hook bolt fastenings. By the end of the	Obvious health risks from fibre release. Friable surface, and very fragile – never walk on such a covering without crawling boards. Corrosion of hook bolts will cause sheeting to split. Often coated with bitumen or rubber solutions as a remedial treatment. Be very cautious of the effectiveness of these treatments.

Materials and defects

	Item, element or material	Effects
	1970s insulation was being added to the roof construction and this brought about condensation problems.	 Rigid foam spacers were sometimes used below cement fibre sheeting in order to create a void into which insulation could be placed. The rigid foam compresses with time and occasional traffic loads, leading to 'chattering' of the roof sheets and possible water ingress.
12	**External ceramic tiling (O)** See 'mosaic tesserae'. Tiles were often prism shaped or ridged in some way.	Similar problems to mosaic tesserae in terms of delamination of background materials.
13	**Flat concrete floor slabs. (Plate floors) (O,W)** Fairly thin slabs with mushroom head thickening around column heads.	Very high shear stress around column head has been found to be cause of structural failure. Beware of flat slab car park construction – recent major collapse of car park of flat slab construction.
14	**High alumina cement concrete (O,W)** Often used in pre-cast and pre-stressed work rather than insitu work. Mainly (but not exclusively) for roof and floor beams. X and I profiles were common. Some variants for pre-cast factory units (portals and purlins). Sometimes used to form an insitu stitch between two pre-cast beam members or column to beam connections. Developed high early strength. Can have a brownish tinge.	Loses strength with age. Susceptible to chemical attack in damp conditions and contact with gypsum plaster. Strength and durability assessments to be recommended. Since 1974 the material should not have been used in buildings. Considered to be a deleterious material. Several roof failures in the 1970s, but no known cases of flooring collapse. 

Materials and defects

	Item, element or material	Effects
15	**Hollow clay pot floors (O)** Quite common during the 1960s. Concrete poured between pots and in the form of a topping. Sometimes screed could be structural. In other case, non-structural screeds may have been removed to gain additional load or headroom. 	Watch out for clay spacer tiles between the pots. These can conceal honeycombing of the concrete rib, lack of fire protection, durability or strength. Removal of tiles and Gunnite repairs may be necessary. 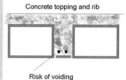
16	**Lack of movement control joints (A)** Cement masonry walls are prone to thermal and moisture movements. Cement mortar is less flexible than older lime mortars, and the stresses induced by thermal movement are relieved by cracking. Centres of joints depend on nature of brick, size and shape of panel etc. Lack of joints was common in 1960s buildings.	Oversailing of brickwork on dpc, particularly in warehouses. Also watch out for joints that have been filled with Flexcell impregnated board, as this is not very compressible and can reduce the benefit of a joint. Early sealants were also resinous and could lead to staining of adjoining surfaces. Hardening and embrittlement of joint sealants to be expected now.
17	**Large panel buildings (H)** A number of different systems were constructed. Large panels formed the external enclosure and also supported pre-cast floor planks or slabs. Connection details were made on site with wire hoops and in-situ work. The design of joints in panel systems was critical in the success or failure of the system. A variety of types were often employed, in some cases using baffles in the form of open drained joints or face sealed joints using mastic or neoprene gaskets. The success of the building from a structural point of view relies on the connection between the individual panels The ability to withstand local damage by means of alternative load paths is critical.	Risk of disproportionate collapse - see Ronan Point disaster. Following this, high rise blocks were strengthened. Some low-rise blocks may not have been checked. Poor quality control of structural connections led to weakness, poor fire stopping or corrosion risk. Possible lack of tying in of precast components. See 'Tying-in' later. Large panel systems suffer from many of the defects described previously; more important rain penetration, corrosion of reinforcement, poor thermal insulation, distortion or physical damage to panels. Calcium chloride was not used in most pre-cast systems although in some types it was added on a batch by batch basis to aid manufacture

Materials and defects

	Item, element or material	Effects
		perhaps where there were particular programming problems. Thus the inclusion of the material is unpredictable.
		The most common faults with these types of systems relates to the gradual deterioration of either the baffle, particularly where butyl rubber was used, ageing of the sealants or ageing of gaskets in face sealed joints. Misplacement of baffles or insufficiently sized baffles in wide joints can lead to water penetration.
		It is quite common to find that quality control standards during manufacture were not up to scratch, with the result that reinforcement was commonly misplaced in the fabrication of the concrete panels. This later led to corrosion of the reinforcement and spalling of the concrete.
		In many cases dry packing used between infill concrete and the panel above is missing or poorly compacted, so that vertical loads are transferred only on the bolt fastenings, resulting in localised cracking around fixing positions.
		In the long term, panels can distort generally as a result of shrinkage in the concrete structure behind and as result of normal thermal movements in the building as a whole. This can lead to damage at joints, displacement of seals and baffles and subsequent water penetration. Furthermore, smoke stopping between floors and compartments can be damaged with the result that in a fire, smoke can transfer rapidly between occupancies or zones.
18	**Mineralite render (O)** A thin (2-5mm) coating of fine grained minerals with a textured surface. Often applied to exposed concrete columns and beams or in larger areas such as spandrel panels. Variety of colours available.	Beware of adhesion failure - can be widespread. Difficult to match repairs.

Materials and defects

	Item, element or material	Effects
19	**Mosaic tesserae (O)** A common finish comprising small (25mm square) ceramic or glass tiles applied to a render background. Often supplied in paper backed sheets of around 300x300 to facilitate laying. 	Adhesion failure of tiles leads to individual tiles falling off and scattering over a wide area. Obvious health and safety risks. More often than not however, it is the render background that fails, losing adhesion to the concrete or brick substrate. This is potentially more serious as larger and heavier sections could collapse. A hammer survey is to be recommended to check for soundness and to identify hollow areas. Repairs are possible using vacuum injection techniques, but hacking off and repair of spalled areas can lead to peel back and cracking of adjoining surfaces and deterioration due to water ingress and freeze/thaw cycles.
20	**Mosaic tesserae in overhead situations (O)** Often used as a soffite finish to projecting balconies, shopping mall covered ways and the like. Tiles would be bedded on render or possibly applied over expanded metal lathing. 	Watch out for corrosion of the metal lathing or fixing screws as these may not have been protected against corrosion. Timber fixing battens behind the lathing can also be a problem. In severe cases, large sections of render and tile finish can collapse. If water penetration is suspected recommend further intrusive investigation.
21	**No fines concrete (H)** Used in the manufacture of large panels for housing and similar structures, intended to create slightly better insulation properties.	Very low level of resistance to carbonation, hence risk of carbonation and corrosion.
22	**Panel joints (O,H)** Panel joints in large panel systems often comprised a plastic or metal baffle sprung into grooves in the edge of each panel. To prevent leakage, it was common to provide a tape back seal to the rear face of the joint.	Baffles may be missing or dislodged. Back seals often missed with consequent risk of water penetration. Flat roof abutments often dressed under the bottom edge of panels, which makes them difficult to repair. Often the need to modify the drained joint into a face sealed joint.

Materials and defects

	Item, element or material	Effects
23	**Reconstituted stone (A)** Often used as window cills or window surrounds, string courses or other projecting features in all types of buildings. Sometimes fixed with ferrous cramps rather than phosphor bronze. Contain light reinforcement.	Propensity to carbonate fairly rapidly with the result that reinforcement corrodes, causing the features to spall. Corrosion of cramps can lead to displacement of features such as projecting window surrounds.
24.	**Render backgrounds (A)** Used in conjunction with tile finishes and mosaic tesserae. Often very strong Portland cement based mixes were used.	Possible adhesion failures on concrete due to presence of traces of mould oil on the surface. Sometimes used water based bonding agents (giving a white milky appearance when render is removed) when there was a poor mechanical key. Later bonding agents were of SBR, which were more durable. Inflexible renders, high vapour resistance and risk of water entrapment.
25	**Reinforced aerated autoclaved planks** Often used as roof decks - 'Sipporex' or 'Durox' or sometimes as vertical walling. Thin reinforcement, 300-750mm width. Made from a Mixture of cement, blast furnace slag, pfa plus aluminium.	If designed before 1980 may deflect excessively, evidenced by transverse cracking on soffite. Some concerns over durability of reinforcement.
26	**Sand faced fletton bricks (A)** These were a popular and cheap brick type manufactured by the London Brick Company near Peterborough. Often found in 1960s housing or industrial applications. Often, but not always, a red/pink colour with a heavy textured wire cut type of surface. Rear face of brick is smooth with colour bands or 'kiss marks' arising from the burning process.	No problem in sheltered applications, but bricks in exposed situations such as parapets, chimney stacks and free-standing walls, where saturation is common, are at risk of sulphate attack. The bricks have a very high sulphate content. When wet, soluble sulphates react with Portland cement bedding mortars, causing the mortar to expand and so disrupting the brickwork. Once this occurs, the damage is terminal. Often rendered in mistaken belief that this will cure the problem, but this is a very short-lived solution and will only make matters worse.
27	**Softwood joinery (A)** External joinery was often of poorly seasoned sapwood with a low life expectancy.	Very poor durability, especially glazing beads, cills and horizontal rails. Further decay where timber has been pieced-in during repair.

Materials and defects

	Item, element or material	Effects
28	**Steel windows and cladding (O,W)** Typical single glazed windows were manufactured using a section known as W20 by Crittal Windows. Either casements or tilt and turn varieties. Larger curtain walled sections were manufactured by coupling window units together with galvanised steel tee bars.	By now, early windows may be paint bound or distorted. Ironmongery may be defective. Timber sub frames were common and may be decayed.
29	**Stramit roof decking (A)** An insulation board often used as a roof decking. It comprises a rigid board about 50mm thickness of compressed straw sandwiched between two layers of building paper. The boards were about 1200x450 width and had a brown paper finish. Check in plant rooms, lift motor rooms, roof access housings and the like. Used in some domestic applications.	The material had a very low resistance to water and would decay easily. For that reason it is less usual to find it nowadays. The boards had a grain and needed to be laid correctly, with support perpendicular to the grain. Failure to do this could lead to distortion of the board and subsequent 'wave' effects in the roof line. This in turn could stress the covering and cause failure. Saturated Stramit board would turn into a brown silage-like mess. Beware of safety issues (risk of collapse) when walking on Stramit roofs.  Failed Stramit roof deck
30	**System built housing (H)** In general terms these types of buildings were prefabricated, based on either steel or concrete construction. Examples would be the British Iron and Steel Federation properties or, if concrete, Woollaway, Unity, Airey etc. The form of construction was generally based upon the erection of a frame with cladding fitted to it or alternatively a panel system.	Problems have occurred as a result of carbonation in the concrete and initial lack of cover, use of unsatisfactory materials such as thin steel tube used as reinforcement, problems of interstitial condensation, damage to sealants etc. In some cases the decay of structural parts has reached severe proportions and it has been necessary to contrive methods of reinforcing the frame or alternatively providing a new cladding system, possibly based on conventional brick

Materials and defects

	Item, element or material	Effects
		and block cavity walling systems.
		In summary, the major problem areas are as follows: -
		Corrosion of reinforcement due to carbonation and/or chlorides, resulting in expansion of the steel and disruption of the concrete.
		Corrosion of metal ties securing cladding panels and the resultant expansion disrupting the cladding and compromising weathertightness.
		Poor or uneven insulation, cold bridging giving rise to interstitial condensation.
		Insufficient fire protection.
		Poor wind bracing.
		Deterioration of timber components such as window frames, particularly where they form part of the primary structure.
31	**Trussed roof construction (H, O)** Trusses were introduced into the United Kingdom during the mid-1960s and were primarily intended for the housing market, although gradual improvements in stress grading and timber engineering have now taken them into commercial, educational and leisure buildings. Commonly designed to pitches of between 20° - 35°, with a span of around 3 - 10m, trussed rafters are usually jointed with factory fixed galvanised steel fasteners, although plywood gussets are sometimes used. With the use of stress graded timber, sizes can be reduced to as little as 35mm in width, with trusses arranged at 450mm or more, usually 600mm centres. Spans of greater than 10m can be achieved, although buckling of compression members can become a problem – and transport to site may be uneconomic.	The correct positioning of the connector plates is essential so that sufficient teeth engage in the timber to prevent the joint, particularly at the apex, from pulling apart. If this happens the truss will settle, or fail. The signs of this may be hogging in the roof and damage to internal finishes.
		Shrinkage of timber after fabrication can affect the adequacy of the truss as a whole. If the timber members cannot meet, all the joint forces will be taken up by the metal connections, which could buckle or pull out. Diagonal bracing was not a requirement of the building regulations before 1976.
		Corrosion of gang nails used in the manufacture of trussed rafters due to interaction with certain timber preservatives.
		Compression members may be subject to sideways buckling under load. Bracing is also needed to prevent buckling.
		Trussed rafter roofs are often clad with large interlocking

Materials and defects

	Item, element or material	Effects
	[Diagrams labelled TR-0 (1 items), TR-1 (1 items), TR-2 (1 items), TR-3 (1 items)]	roof tiles which, together with the tiling battens, tend to produce a relatively stiff plate. They are none the less vulnerable to long term vibrations which can lead, in an inadequately braced roof, to lateral buckling where the trusses adopt a sideways lean - domino effect - either in one or two directions towards gable walls. The structure will also be expected to afford support to gable walls and this is usually achieved by the use of galvanised steel straps turned down into the cavity and fixed to at least two adjacent trusses. Lack of adequate restraint to gable walls or other unrestrained elements could allow an unacceptable degree of movement to take place.
32	**Tying-in of precast concrete floor and roof slabs (O,H)** Prior to 1972 (CP110) tying-in was left to engineering judgement. There is a need to form a connection between wall structures and internal precast floor planks. This can be achieved with continuous metal straps or structural toppings to prevent planks from gradually moving apart.	Failure to tie-in properly can lead to the elevation gradually parting company from the floors. Evidenced by a series of parallel cracks in the floors, gradually increasing in severity higher up the building. If neglected, collapse could occur under accidental loads.
33	**Vitrilite panels (O)** Used in conjunction with steel windows as above, these single glazed spandrel panels were made from annealed glass with a powder coating fused into the rear surface during manufacture.	Risk of failure due to heat build up in spandrel panels, bird strikes or mechanical damage from cradles. Water penetration can cause staining and deterioration of rear surfaces. Replacement panels no longer available and may have been made from painted glass with less life expectancy.
34	**Wall ties (A)** Often wire butterfly ties. Thin steel sections and poor galvanising standards. Cavity walls were rarely insulated and cavity tray detailing may be poor. During the 1960s it was common to use galvanised wire ties and, in some cases, vertical	Factors which could have an influence on the life of the tie are the steel alloy used, the quality of the protective coating and the mortar type - particularly if this was contaminated with chlorides or if the building was in an exposed location. Research work undertaken by

Materials and defects

	Item, element or material	Effects
	twist ties with sub-standard protection coatings. The life expectancy of bitumen and zinc coatings on these ties is frequently well under the 60 years that was originally predicted. In fact, in 1981, BS 1243 tripled the minimum allowed zinc coating thickness on wire ties.	the Building Research Establishment suggests that average zinc loss is about 2.1 microns a year. For pre-1981 ties, this results in a predicted coating life of 12 – 26 years for wire ties and 25 – 46 years for vertical twist ties. On the inner leaf, where the circumstances are less aggressive, the zinc coating can be expected to last much longer.
		If wire ties have been used, these have the unpleasant tendency of corroding away without any substantial physical disruption to the brickwork. Damage becomes manifest by the sudden collapse of an outer leaf particularly in conditions of high wind.
		With the thicker, vertical twist ties the amount of metal is significantly more and if corrosion occurs it is likely that the thickness could increase by as much as four times. The cumulative effect of this corrosion will be the creation of horizontal cracks in the brickwork and eventually the lifting of the roof covering at eaves level to give the so-called pagoda effect.
35	**Woodwool as permanent shuttering (W,O)** Often used in basement carparks where additional insulation was required, or in some office buildings.	Risk of poor compaction of concrete during placing, or grout loss leading to honeycombing around re-bars. This could prejudice fire protection, durability or in extreme cases strength. Intrusive investigation required to determine if steel is covered properly.
36	**Woodwool slab roof decks** Often with galvanised steel tongue and grooved edge strips and with a pre-screeded finish, or an applied finish reinforced with chicken wire. Size about 1200x 450 or 600mm. (See for use in permanent shuttering). The material offered some thermal insulation qualities. Often used	Reasonably durable and, contrary to popular belief, does not degrade rapidly when wet. However, failure of screed is probable during re-roofing operations, leading to need to renew the deck.

Materials and defects

	Item, element or material	Effects
	in plant room roofs, access housings and the like.	
37	**Cladding systems (O, W)** Early curtain walling systems relied upon the use of galvanised steel window components coupled together or fixed within framed openings. These systems were often single glazed and incorporated Vitrolite spandrel panels. In the early 1970s more aluminium curtain walling systems were developed. Early systems were single glazed and face-sealed but, latterly, drained systems were installed, incorporating double glazed units.	Inspect opening sashes for signs of distortion - usually due to paint build up. Window fittings are usually worn or inoperable. Pay particular attention to the security of fanlight fixings. Early double glazed systems were fully bedded. Volatalisation of sealants is common, leading to voiding, leakage and deterioration of edge seals. Be suspicious of early face-sealed systems in terms of future durability. 
38	**Concrete frame** Expressed concrete frames were common in the 1960s, often with brick infill panels. In the 1970s there was a move away from this form of construction to brick cladding, with the frame concealed either by brick slips or by a brick outer leaf supported on steel angles.	In both cases, there is a risk that the brick panels can become stressed as a result of the normal shrinkage (axial forshortening) of the concrete frame. A failure to provide movement joints means that loads can be transferred to the panels with the result shown in exaggerated terms below.

Materials and defects

Defects in concrete

The failure of concrete is often revealed by cracking, spalling and corroded reinforcement. While the outward symptoms of a number of faults may appear similar, repair methods must be based on a sound analysis of the cause.

Cracking can be caused before hardening due to workmanship problems or after hardening due to physical, chemical, thermal or structural effects.

The alkaline nature of concrete protects steel reinforcement against corrosion, a feature termed 'passivity'. A reduction in the level of alkalinity gives a risk of corrosion in steelwork provided water and oxygen are present. A summary of common defects and consequences are identified in the following table.

Fault	Reasons	Results
Carbonation	Carbonation is generally of concern on exposed concrete surfaces and will take place very slowly with high quality dense concrete. Features such as poorly compacted concrete, cracks or other fissures in the surface and the type of aggregates used could all affect the rate of carbonation.	A reduction in passivity of concrete coupled with water and oxygen can lead to corrosion of steel reinforcement and subsequent spalling of concrete. Lack of sufficient cover to reinforcement may allow steel to fall within the carbonated layer and become at risk. It has been found that the depth of carbonation is roughly proportional to the square root of the time $d=k\sqrt{t}$ (d=depth, k=constant, t=time). The constant will vary with the properties of concrete. By examining the age of the concrete and the depth of carbonation it is possible to make a rough prediction of the depth of carbonation at a future date.
Rust stains	An indication of reinforcement corrosion but avoid confusion with staining ferrous sulphide inclusions in the aggregate or rusting of small diameter wires. Chlorine attack is difficult to deal with effectively and needs careful diagnosis (see later). Chloride induced corrosion will often result in dark stains around corroded reinforcement.	Although unsightly, staining such as this is unlikely to indicate reinforcement corrosion unless cracking is visible.

Materials and defects

Fault	Reasons	Results
Chloride attack		See separate section on chlorides on page 163.
Sulphate attack	Sulphates are present in varying levels in many substances including gypsum plaster, certain aggregates and sub-soils.	Possible disturbance to foundations and walls due to expansion of floor slabs. Loss of strength, material becomes friable. Rate of carbonation increases, expansive reaction occurs in the concrete.
Alkali aggregate reactions principally Alkali Silica reaction	As a result of chemical interaction between alkaline fluids in concrete and reactive minerals in certain types of aggregates, a calcium alkali silicate gel is formed. This gel takes in water and expands.	Relatively rare fault in UK structures. In non reinforced concrete this cracking often illustrates a random pattern of fine, almost invisible cracks bounded by some larger cracks. This cracking is easily confused with shrinkage cracking or even frost attack. Gel can sometimes be seen on the surface of the concrete, possibly coupled with spalling lenses.
Aggregate reactions	Alkali carbonate reaction and alkali silicate reactions have similar problems, but are much less common in the UK.	In reinforced concrete, cracks tend to run parallel with reinforcing bars or prestressing tendons. In particularly severe cases, gel may be visible. Microscopic examination of concrete is the only sure way of identifying attack. Although rare, ASR is often only found in structures exposed to water.

Fungi and timber infestation in the UK

Fungi and moulds

Fungi live on dead organic material and play a natural role in the breakdown of dead organic material, which includes timber. Most timber is too dry for fungal growth but timber decay can occur if the moisture content (MC) is increased.

Sporophores (fruiting bodies) are often the first indication of a problem.

Wood rotting fungi familiarly is classified as dry and wet rot. Serpula lacrymans is the only true dry rot – there are many wet rots. Serpula lacrymans can decay timber at a much lower MC than any of the wet rot fungi and can penetrate masonry/brickwork and behind plaster.

The accompanying table should aid identification of various fungi.

Materials and defects

Fungi identification guide

Type	Usually found	Effect on timber
Wood rotting fungi		
Dry rot Serpula cuboidal	Inside buildings, mines, boats – never attacks timber outside	Large cuboidal cracking (brown rot)
Wet rots Conrophora Puteana (cellar fungus)	Most common of wet rots in buildings. Associated with serious leaks – fouled plumbing etc. Also decays exterior	Cuboidal cracking – small cubes (brown rot). May leave thin veneer of sound timber. Affected wood becomes dark brown
Fibrioporia Vaillanti (mine or pore fungus)	Associated with water leaks. Most common species of poria group	Cubodial cracking – large cubes (brown rot) Affected wood darkens
Phellinus Contiguous	Decay of external joinery (softwood)	Timber becomes soft. (a white rot) Wood becomes fibrous
Phellinus Megaloporous	Attacks oak heartwood. Presence often associated with death-watch beetle	
Corioius Versicolor (Polystictus)	Most common white rot decay of external hardwood	No splitting or decay but much weight loss
Lentinus Lepideus (Stag's Horn fungus)	Rare, but sometimes in flat roofs	Cuboidal cracking Darkens woods. Wood feels sticky
Non-wood rotting fungi		
Peziza (Elf-Cup)	Occurs on saturated masonry or plaster, internally and externally. Associated with leaks	
Moulds		
Aspergillus Penicillium Pullularia	Almost any damp surface in humid conditions	Superficial – easily removed

Materials and defects

Mycelium	Fruiting body	Conditions for growth
Cotton wool-like if damp. Greyish white with purple/yellow and lilac patches if dry	Reddish brown centre, white margins. Flat plate or bracket shape. Possibly red spore dust nearby	Timber MC 20-40% (slightly damp) Temperature 0-26°C
Brown branching strands on wood and masonry or brickwork. Usually not in daylight areas	Rarely found inside. Flat plate-like. Greenish brown centre, yellow margin. Knobbly surface	Timber MC 45-60% (very damp) Temperature -30°C-+40°C
Strands flexible when dry. White	Plate shaped – white pores. Rare	Timber MC 45-60% (very damp) Temperature up to 35°C
Light brown masses	Plate-like with pores. Dull brown	Timber MC 22%+ Temperature 0-31°C
Yellow	Large, plate-like, hard. Various browns in colour	Timber MC 20-35% Temperature 20-35°C
Rarely seen	Up to 25mm across. Hairy ringed zones to pores to underside	
Soft whitish needle shaped crystals on surface	Some resemble stags horns, others are inverted mushrooms on stalk – brown	Timber MC 26-44% Temperature 25-37°C
	Buff coloured and fleshy. Distinctive	
Like coconut matting	Toadstool – white head – spores released in black ink type liquid. Microscopic but spores show up as various colours: black, green, white, brown, yellow, pink	Very humid conditions

Materials and defects

Various superficial coloured moulds, often seen affecting timber to buildings, are easily removed, but indicate a MC that might permit more serious fungal decay. Lowering the ambient humidity by increasing ventilation is generally the best course of action. The presence of non-wood rotting fungi also suggests conditions suitable for dry or wet rot.

Treatment

There are many specialist timber treatment companies who will carry out surveys, analyse samples and guarantee any eradication work they undertake.

Briefly, the traditional specification for treatment of dry rot will include the following:

- ❖ identify cause(s) of dampness and effect cure;
- ❖ cut out timber to 0.5m beyond decayed wood, remove and burn;
- ❖ hack off plaster, rendering and remove skirtings, architraves and other joinery from area to be treated to 1.0m beyond infection;
- ❖ remove surface mycelium from exposed masonry and wire brush;
- ❖ surface spray exposed masonry with fungicidal wall solution at manufacturer's recommended rate of application;
- ❖ consideration should be given to irrigating masonry, although it is unlikely that this treatment will be as effective as one might hope;
- ❖ replacement timber to be treated to BS 5268, 1977 and thereafter, treat with preservative to BS 5707; and
- ❖ existing timber to be cleaned and sprayed with organic solvent preservative and further treated by application of timber paste.

Treatment of wet rot is similar, although affected masonry need only be isolated from the source of the dampness. In some circumstances chemical treatments can be minimised and the outbreak controlled by environmental manipulation, but this may not always be practicable. Where chemicals are used, request specialists to provide COSHH assessments. Permitted fungicides include zinc acypetacs and tri (hexylene glycol) bioborate. All specified treatments should display an HSE number.

In recent years there has been a trend towards less intrusive forms of repair. If the source of moisture can be removed, the fungus will die. Thus, by managing the building environment the problem of decay can be minimised, and the extent of disruption and intrusive surgery kept to a minimum.

The importance of safety measures cannot be over emphasised and it is vital that current safety legislation is understood and complied with, including the Control of Pesticides Regulations Act 1986 and Control of Substances Hazardous to Health Regulations Act 1988.

Under the Wildlife and Countryside Act 1981, it is an offence to spray roof spaces that may harbour bats without the approval of the Nature Conservancy Council.

Insect infestation

In this country wood boring beetles are the major group of timber attacking insects. The life cycle of the various beetle species is similar; adult female beetles lay eggs in cracks or the end grain of the timber, larvae emerge from the eggs and feed on the wood leaving frass in the tunnels

Materials and defects

they bore. The larvae enter the pupal stage and soon after emerge as adults leaving flight holes in infested timber. Some species cause only superficial damage.

Examples of most common wood boring beetle species are as follows:-

- ❖ Common furniture beetle: very common, estimated that up to 80% of houses over 40 years old in rural areas are affected. Infestation often in damp areas of house, for example, beneath WC. Flight holes 1.5-2.0mm diameter. Adult beetles emerge May-September.
- ❖ Death-watch beetle: infestation uncommon, often found in ancient buildings and therefore more expensive to eradicate. Confined to south and central parts of England and Wales. Attacks elm, chestnut and oak. Adult beetles emerge in Spring through flight holes up to 3mm wide. Presence indicates fungal attack.
- ❖ Bark borers: found in timber where bark not completely removed. Larvae confined to bark areas and hence damage caused is superficial.
- ❖ House longhorn beetle: only found in Surrey, Berkshire and Hampshire. Regulations require new timber to be treated prior to use.
- ❖ Wood boring weevils: several species – only attack partially decayed timber, cause considerable damage.
- ❖ Powder post beetle: few flight holes, convert timber to powder leaving veneer of 'sound' timber.

Troublesome plant growth: Japanese Knotweed

Introduction

Phillip Franz Balthasar Von Siebold, a Bavarian physician in the Dutch East Indies army, introduced Japanese Knotweed to Europe. Since its introduction into Victorian gardens in the nineteenth century, it has been spreading into both wild and cultivated areas. It is now present in almost all of the 3,880 10km grid squares that cover the UK.

Key facts

- ❖ Dioecious plant.
- ❖ Rapidly colonising riverbanks and areas of wasteland.
- ❖ In the spring, it can grow to a height of approximately 3 metres.
- ❖ In mid-summer it grows with stiff, bamboo-like stems. During this period, the plant produces large masses of white flowers, but it is not thought to produce viable seeds.
- ❖ In autumn, the leaves and stems die, but remain stiff and erect.

What does it look like?

Japanese Knotweed is a plant with spreading rhizomes and numerous reddish-brown, freely branched stems. The plant can reach four to eight feet in height and is often shrubby. The leaves are four to six inches long and generally grow with an abrupt point. Male and female versions of the inconspicuous flowers are produced on separate plants. The approximately 3mm fruits are brown, shiny and triangular.

How is it spread?

- Extremely rare for the seeds of this plant to germinate.
- The most common method of dispersal is by means of stem, crown and rhizome (underground stem) sections.
- The spread of this plant to new sites is frequently due to the transportation of contaminated topsoil, because fragments of rhizome as small as one gram can produce new plants.
- Spread along riverbanks may also occur when the shoots are cut and allowed to float downstream and are washed onto damp soil beside the river.

Specific problems caused by Japanese Knotweed

- Damage to paving and tarmacadam areas.
- Damage to flood defence structures.
- Damage to archaeological sites.
- Reduction in land values.
- Increased risk of flooding through dead stems washed into the river and stream channels.
- Aesthetically displeasing.

Control methods

Mechanical control

Digging, ploughing and dredging are unlikely to have any long-term benefit because of the extensive nature of the rhizome system and because of the ability of even small fragments to re-grow.

A disadvantage to flail mowing that should be considered is that it is highly likely to spread the stem material, allowing new growth in non-infested areas.

The control named above is a short-term measure, and will not eradicate this plant.

Chemical control

The only herbicide approved for use in or near water that controls this plant is glyphosate. The green shoots should be sprayed with glyphosate, when the plant is about a metre high, usually in May. Total eradication is unlikely following a single application and to achieve this, shoots should be cut at least three weeks after the initial spraying so that any re-growth can be sprayed later in the season, usually in July or August.

Biological and environmental control

To date there is no form of biological control known. Furthermore, once the plant has become established, there is no environmental method of controlling it.

The Law

The Wildlife and Countryside Act, 1981, made it an offence to spread Japanese Knotweed. Section 34 of the Environmental Protection Act, 1990, places a duty of care on all waste producers to ensure that any wastes, and any specific harmful properties, is provided to the site operator.

Any excavated soil from areas where Knotweed has established must be disposed of at a licensed landfill site and not reused in further construction or landscaping.

Cost implications

- ❖ The recent British government review of non-native species policy estimated the cost of controlling Knotweed countrywide at £1.56 billion; this gives an indication of the extent of the problem and the high costs associated with control were it to be attempted.
- ❖ Chemical management would be the most effective method of control, although this could cost up to £10/m^2, if relandscaping costs are included.
- ❖ Disposal rates of £50/tonne, combined with the difficulty of finding a site that will accept it, make alternative methods equally costly.

Rising damp

Research by the BRE and others suggests that rising dampness is often misdiagnosed by surveyors and so-called damp specialists, with the result that costly and unnecessary remedial treatments are specified.

In many cases, diagnosis is undertaken by means of electrical resistance or capacitance meters, but these can give very misleading and unreliable results in materials other than timber. Surveyors should not diagnose rising damp without first having undertaken a proper study. When rising dampness is suspected, do not automatically recommend a specialist inspection – more often than not the specialist will use exactly the same resistance equipment to make his diagnosis. A more reliable method has been prepared by the BRE and may be found in BRE Digest 245.

Symptoms

Typically these may include:

- ❖ damp patches;
- ❖ peeling and blistering of wall finishes;
- ❖ a tide mark 1m or so above floor level;
- ❖ sulphate action;
- ❖ corrosion of metals for example, edge beads;
- ❖ musty smells;
- ❖ condensation; and
- ❖ rotting of timber.

The above symptoms do not of themselves indicate the cause of dampness. Common causes could be lateral rain penetration, condensation or entrapped moisture. High external ground levels, bridging of damp proof courses, defective rainwater goods and the like should all be self-evident and could give rise to similar symptoms.

Soluble salts are present in many building materials. The salts can be dissolved and moved to the surface of the element as evaporation takes place. Hygroscopic salts (typically, nitrates and chlorides from groundwater) can be present in some materials. These salts absorb moisture from the atmosphere, and can in certain circumstances cause extensive staining and disruption of finishes.

Other possible sources of salt contamination include chemical spillage, splashing from road salt etc.

Materials and defects

Equilibrium moisture content

Many building materials absorb moisture, and when exposed to damp air will attain an equilibrium moisture content. This hygroscopic moisture content (HMC) will vary according to relative humidity. Typical relationship curves can be plotted for different materials, and although these can only establish general indications it is possible to compare readings from different materials in the construction of a wall. For example at 75%RH the MC of yellow pine would be 13% while the MC of lime mortar would be 2% and 0.5% in brick.

Some materials possess an HMC of as much as 5% without the introduction of salts from external sources. This figure should be regarded as an appropriate threshold as to whether or not remedial action is likely to be required.

Measurement of moisture content

Resistance or capacitance meters can give misleading results.

The presence of soluble salts on the surface of a wall will cause an electrical resistance meter to indicate a high reading, even if the wall were otherwise dry. Deep wall probes may give a more accurate picture, but will still be affected by soluble salts, as these are generally highly conductive.

A Speedy Moisture Meter will give a much more accurate reading of MC in all materials. (Resistance meters are usually calibrated for use in timber and can give an approximation of MC in that material).

The Speedy meter comprises an aluminium flask fitted with a pressure gauge and a removable lid. Using a 9mm drill on slow speed, a sample of dust is taken from the brick, mortar or plaster. The sample is weighed and placed into the flask. A small quantity of carbide is then added and the flask sealed. Moisture in the sample reacts with the carbide to form acetylene gas. The pressure of that gas is then read off the pressure gauge, which is calibrated to read %MC. With care, the meter can give a very accurate reading, comparable with laboratory kiln dried tests.

Rising damp

Rising dampness within a wall is in a sensitive equilibrium. There must be a supply of water at the base of a wall and the height to which that water will rise depends upon the pore structure, the brick, plaster or other finish. Water will also evaporate from the surface of the wall at a rate dependant upon temperature and humidity.

During wet weather, evaporation may decrease and ground water tables may rise, giving rise to an increase in the severity of the dampness. The reverse may happen during dry spells, and evaporation will be increased by central heating.

Soluble salts derived from groundwater or building materials will complicate the situation. Salts will increase the surface tension of the water and so draw it further up a wall. Furthermore, as evaporation occurs, stronger salt solutions are drawn towards the surface and may eventually crystallise out. This process reduces the amount of evaporation and so may raise the height of the dampness. The soluble salts are often hygroscopic and absorb moisture from the atmosphere. If this occurs, the situation will appear worse during wet weather and better during dry.

As noted above, the presence of hygroscopic salts does not necessarily indicate rising dampness.

Diagnosis of rising damp

BRE Digest 245 sets out a method of diagnosis. The method involves drilling samples from the wall to measure both their moisture content and hygroscopicity (HMC). Samples are taken from mortar joints from 10mm to a depth of 80mm every two or three courses from floor level up to a level beyond that which damp is suspected. While the carbide meter can be used to measure moisture content (MC) it will be necessary to send samples to a laboratory to measure HMC – to see if the samples could have absorbed the quantity of moisture found from the atmosphere.

By subtracting HMC from total MC, it is possible to determine the value of 'excess' moisture, which could result from capillary action or water from other sources. The comparison of HMC and MC gives an indication as to which is controlling the dampness at any position. If MC is greater than HMC, then moisture is coming from some other source. If the reverse applies, then moisture is coming from the air. Plotting the results graphically can then assist in gaining an accurate picture of what is happening.

Surface damage arising from hygroscopic salts can be significant. The HMC of contaminated wallpaper or plaster can be as much as 20%. It follows therefore that contaminated plaster will need to be removed. BS 6576 deals with this subject in more detail.

Rising groundwater

During the latter part of this century, the level of ground water beneath major UK cities has risen rapidly, leading to increasing concern that huge costs could be incurred from damage to buildings and infrastructure if preventative measures are not taken. The problem stems from the city-centre industries that populated the areas during the industrial revolution and their demand for water. Beneath London this led to a reduction in the level of up to 90 metres. The usage has declined significantly since the late 1960s as the industries relocated and without this extraction levels have recovered, rising by 1.5 metres a year initially and by as much as 3 metres a year recently. By the late 1990s, water levels in central London had recovered by 35 metres, close to the 1900 level.

It was therefore realised that action to minimise the damage was needed urgently to allow time for planning and implementation. A group was formed from interested parties and in March 1999 it announced that a five-stage plan had been developed to safeguard London from the effects of rising groundwater. The plan involved controlled increased abstraction from 50 or more existing and new boreholes, with the amount that could be used for drinking water maximised and the remainder to be used for industrial or agricultural processes. The five phases of the strategy are:-

- Utilise four existing licensed water supply boreholes on the outskirts of London.
- Equip three proven borehole sites near central London with the latest filtration technology to provide further water for drinking purposes.
- Encourage the use of existing and new private boreholes, in and around central London where the water quality is lower, for non-potable uses.
- Identify locations for new control boreholes in central London, initially to be pumped to waste until end uses can be determined.
- New control boreholes in outer London to intercept the flow of water towards the central basin.

The plan has the backing of government and is being managed by Thames Water, working in conjunction with the Environment Agency. Although the threat to London is the most immediate, other cities are at risk and it is likely that similar strategies will be implemented as required.

Materials and defects

Subsidence

Many of the UK's housing built on what are known as shrinkable clays have quite shallow foundations, usually less than 1m deep. Such clays are generally strong and able to support a building of four storeys on a single strip or trench-fill foundation. These soils shrink when their moisture content decreases and then swell when it increases. Slight movement of houses on foundations is therefore inevitable as a result of seasonal changes in moisture content resulting in downward movement or subsidence occurring during the summer and upward movement or heave during the winter.

Greater movements may occur during long periods of dry weather and may lead to sticking of doors and windows. Severe movements are almost always associated with localised subsidence caused by trees whose roots extract moisture from the soil. Conversely, removing a tree tends to cause heave as moisture gradually returns to the soil. Large broad-leaved trees of high water demand are notorious for causing damage.

Before World War II, it was common practice to use shallow foundations no more than 0.45m deep. Houses built within the past 25 years should comply with guidelines issued by the National House Building Council [NHBC] and the British Standards Institution [BSI]. The former requires foundation depths often well in excess of 1m, while the latter requires a minimum depth of 0.9m for any buildings founded on clay and deeper foundations where there are trees nearby.

Because of the link between clay shrinkage and the weather, insurance claims for subsidence damage increase in long dry periods. There is an upward trend in claims and presently the background level of subsidence and heave damage in the UK is about £350M annually. Analysis of insurance claims indicates that of those cases involving foundation movements caused by the shrinkage or expansion of clay soils, 75-80% are exacerbated by moisture abstraction by trees.

Subsidence may also be a consequence of mining activity. The extent of subsidence due to mineral extraction depends on the method used for winning the minerals from the ground, whether by mining, pumping or dredging. The main problems in Great Britain arise from coalmine workings.

In many coal fields in Britain the presence of old workings remain as a constantly recurring problem in foundation design where new structures are to be built over them. If the depth of cover of soil and rock overburden is large, the additional load of the building structure is relatively insignificant and the risk of subsidence due to the new loading is negligible. If however the overburden is thick, and especially if it consists of weak crumbly material, there is a risk that the additional load imposed by the new structure will lead to local subsidence.

The risk of subsidence associated with coal workings may be obtained via a Coal Mining Report obtained from The Coal Authority.

It is important to differentiate between subsidence and settlement.

- ❖ Subsidence is vertical foundation movement caused by failure or shrinkage of the sub-soil.
- ❖ Settlement is vertical foundation movement caused by an increase in applied load.

Testing

Materials and defects

▶ Chemical and physical testing requirements — 196

▶ Non-destructive testing — 197

Materials and defects

Chemical and physical testing requirements

The following are some of the more common requirements for specific tests, although it is better to discuss the quantity or type of sample required with the testing laboratory before sampling to ensure that a suitable and representative sample is provided.

Testing laboratories should be provided with an outline of the location of the sample and the nature of the element inspected. This is often critical in evaluating results and the possible level of risk involved.

Primary rules

- ❖ avoid contamination of sample
- ❖ label clearly
- ❖ inform laboratory of purpose of test
- ❖ consider health and safety aspects when dealing with hazardous substances; seek expert advice first
- ❖ for comparison, 10 grammes = a sugar cube.

Chloride ion content

About 30-50 grams (enough to fill a 35mm film canister) of sample material is required. Drilled dust is preferred obtained using a 10mm percussion drill bit. Discard first 5mm depth of material to avoid contamination from paint, plaster or other surface effects. Take samples from two adjacent holes drilled to the depth of reinforcement. As chloride levels will vary, many samples should be taken for analysis. For chloride profiling to identify ingressed chlorides then take incremental samples at depths typically 5-25, 25-45 and 45-65mm.

Carbonation

Best carried out in-situ. Drill two 10mm diameter holes and break out concrete in between. Treat freshly exposed concrete with phenolphthalein. Concrete will turn pink if un-carbonated. This is not a precise test and carbonation should be recorded to the nearest 5mm.

If you find the depth of carbonation by testing and you know the age of the building, then the rate of carbonation increase in the future can be predicted according to the following formula: $d = k \sqrt{t}$ where d = depth of carbonation, k is a constant and t = age of concrete (years).

Refer to BRE Digest 405.

High Alumina Cement (HAC)

Dust samples taken, as for calcium chloride ion, to determine aluminium content and assess 'proof negative' the presence of HAC. Differential Thermal Analysis (DTA) required on lump samples for 'proof positive' confirmation. Phenolphthalein test for carbonation is not applicable to HAC. It is necessary therefore to remove a lump sample of sufficient size to enable a thin slide to be made for petrographic laboratory analysis. This test is very expensive.

Refer to BRE Digest 392.

Sulphates in concrete

As for calcium chloride ion (30g sample should be sufficient for chloride sulphate and HAC).

Plaster/mortar

To determine mix proportions including, for mortar, cement content. Take dust or preferably solid sample as for chloride ion content.

Asbestos

Air monitoring and bulk sampling by specialists only. Samples must be taken with due regard to health and safety. Seal loose material after sampling. Sample of loose insulation may be taken with corer. About a thumbnail size sample is sufficient for analysis. Laboratory to be UKAS accredited. Seal remaining material after sampling. Laboratory to report on asbestos type, % content and density.

Ground water

Approximately 1,000ml of water should be sufficient to determine chemical signature of ingressed water. A similar quantity of mains tap water should also be taken simultaneously so that the results may be compared. Results of laboratory analysis sometimes are not conclusive.

Leaks may also be identified by 'sounding' by placing listening rods on pipe valves, covers etc. This work is done by specialists often when quiet at night.

Decayed timber

Provide sample large enough to indicate pattern of cracking and/or mycelium growth. Powder material is not sufficient except in cases of insect attack, where frass may give indication of type of beetle.

Non-destructive tests

See BS 1881: Part 201, 'Guide to the use of non-destructive methods of test for hardened concrete' and information supplied below.

Non-destructive testing

There are a range of non-destructive testing (NDT) techniques which may avoid the requirement for large scale costly and disruptive opening-up. Often the findings of non-destructive testing survey may allow opening-up to be targeted in specific locations. Non-destructive testing techniques are particularly useful to establish construction details and the condition of the structure and fabric of a building, in particular relating to building defect analysis.

A summary of the more common NDT techniques are as follows:-

Impulse radar

Radar is a widely accepted non-destructive technique for establishing details of the construction and condition. Radar is effective through most building materials including concrete, asphalt, brick, stone and also through soil. Defects may be mapped without damage or disruption to surface finishes.

Radar is an echo sounding technique where pulses of radio energy are transmitted into the structure by an antennae moved over the surface. Where material boundaries are encountered the different electrical properties reflect part of the energy back to the surface where it is detected by a receiver. Sampling is rapid and collected data effectively forms a continuous cross section enabling rapid assessment of thickness, arrangement and condition over large areas.

Thermography

Infrared thermography is a non-destructive testing method involving precise measurements of surface radiation to reveal changes in thermal performance caused by hidden changes in the construction or in a physical condition. Recent improvements in imaging and processing technology has enabled improved and compact handheld thermographic cameras to undertake surveys of large areas relatively quickly.

Modern thermo-imaging cameras are able to detect variations in surface temperature as small as $0.10°C$. A major limitation of this method is that sufficient heat differential is required between the inside and outside of the building fabric. Environmental conditions favourable for thermographic surveys are therefore restricted to, for example, cold still winter nights when heat flows from the inside to the outside of walls. A major benefit of this technique however is that a great deal of work can be conducted some distance from the subject.

Ultrasonic pulse velocity

Ultrasonic pulse velocity is a non-destructive technique used in testing a wide range of building materials to determine properties including compressive strength and to investigate defects, such as the presence of delamination and the depth of cracking. The method can be applied to materials such as concrete, ceramics, stone and timber. The main advantage of this method is identifying general changes in condition such as areas of weak concrete in a generally sound structure. The technique involves the measurement of the compression wave velocity from a transmitter to a receiver.

NDT investigations

Often one or more of the above techniques may be required. Information may be provided relating to:

- debonding and delamination;
- compaction and voidage;
- spalling or micro-cracking;
- general details of construction, material types and layers;
- moisture content;
- reinforcement or other embedded ferrous metal; and
- brickwork details and conditions.

It is rare that NDT techniques will themselves tell the whole story. It is essential that NDT is considered as an investigation tool as part of a wider engineering assessment.

Cladding

▶	Composite panels	200
▶	Curtain walling systems	201
▶	Mechanisms of water entry	205
▶	Glazing – windows and doors satisfying the Building Regulations	206
▶	Spontaneous glass fracturing	208

Materials and defects

Materials and defects

Composite panels

There has been some speculation of late that the use of metal faced sandwich panels is unsatisfactory, and that while buildings constructed from these materials can satisfy building regulations, there is an increasing reluctance on the part of building insurers to effect cover at reasonable commercial rates.

Given parallel concerns over fire safety and risks to occupiers and fire fighters it is clear that there is considerable potential for bad or misleading advice and unjustified criticism of composite panels in general.

In 1993, two firemen lost their lives tackling a serious blaze at the Sun Valley poultry fire in Hereford. The fatalities were due to the early collapse of plastic foam cored sandwich panels, which contributed to the fire and generated significant quantities of thick black smoke and noxious fumes.

Sandwich panels are of a composite construction comprising two outer layers of steel or aluminium sheet with an inner core of an adhesive bonded lightweight core material. The resulting product is lightweight yet strong and able to span greater distances than would be possible with the individual component parts in isolation. Flexural strength is attained by maintaining the bond between the layers. One side will be in tension, the other in compression. Remove the bond to one face and integrity is destroyed and the panel will fail.

There are broadly four types of core:

- ❖ Foamglass (fairly rare)
- ❖ Polyurethane (PIR or PUR)
- ❖ Polystyrene (EPS)
- ❖ Mineral wool.

If subjected to fire, polyurethane cores (which are thermosetting) will undergo localised charring, although flaming can take place if flammable vapours are released. The charred material will shrink and can lead to delamination of panels.

Polystyrene materials may burn fiercely, give off thick black smoke and allow burning droplets to fall.

Some of the adhesives used in mineral wool products can be combustible.

Foamed glass is generally non-combustible.

The two main issues are:

- ❖ instability of the panel facings in the event of a fire; and
- ❖ combustibility of the panel core material – method statements for fighting fires in composite panel buildings talk about 'fires of the building', rather than fires within the building.

Issue 1 was a contributory factor in the fire at Sun Valley, and could affect composite panels that are not properly supported or restrained. In the case of roof and wall cladding, there are usually additional supports in the form of purlins and sheeting rails and primary fasteners, which serve to tie the two leaves together and prevent them from becoming detached. However, in internal situations such as food processing plants, cold storage facilities etc, the quantity of insulation required will often lead to panels of 200mm thickness or more. Such panels require less support and so present a greater risk of delamination. It is these panels which cause the greatest levels of concern – particularly as the core materials are often no more than EPS.

There will be a conflict between the requirements of the Building Regulations and the requirements of insurers. Building regulations are aimed essentially at ensuring the health of users and neighbours of the building (and of course visitors as emergency services). Insurers may want higher standards of protection, fire suppression and/or more reliance upon non-combustible materials.

Materials and defects

The relevant standards for composite panels were originally set by the Loss Prevention Council but are now administered by BRE certification under LPS 1181. Generally, materials with a PUR or EPS core will almost certainly not satisfy the requirements of LPS 1181, whereas PIR or stonewool products can be engineered to comply.

The Association of British Insurers has produced a report on the issue and it is hoped that insurers will now take a more relaxed view if buildings have been constructed using materials certified under LPS 1181 Part 1 (for external systems) or Part 2 (for internal applications). However, many buildings have been constructed (and still are constructed) using materials that do not satisfy these standards and, in these circumstances, it is important to consider the overall level of risk rather than the mere existence of the panels.

Whether or not sandwich panels constitute a risk is a matter of judgement and scientific fire risk assessment. A reasoned approach may involve the consideration of the following:

- Is there a sprinkler installation in the building?
- Are there any specific fire risks – use, storage of inflammable materials, arson etc?
- Are the panels in the vicinity of battery charging areas?
- Are the panels perforated such that the cores are exposed?
- How are the panels fixed – are they properly restrained?
- What is the nature of the insulant?
- What is the extent of the material and to what extent could it contribute to fire load?

Curtain walling systems

Curtain walling is a weatherproof and self-supporting enclosure of windows and spandrel panels in a light metal framework which is suspended right across the face of a building, being held back to the structure at widely spaced joints.

Types of curtain walling system

Stick system

Stick construction is the traditional form of curtain walling, comprising a grid of mullions and transoms into which various types of glass and/or insulated panels can be fitted. Most of the grid assembly work is done on site. The advantages include relatively low cost and the ability to provide some dimensional adjustment. The disadvantages are that performance is workmanship sensitive. It is not unusual to find systems failing initial waterproofing tests during erection.

Unitised system

Unitised systems comprise narrow width storey height units of aluminium framework containing glazed and/or opaque panels. The entire system is pre-assembled under factory controlled conditions. Mechanical handling is required to position, align and fix units on site onto pre-positioned brackets attached to the floor slab or the structural frame. Modern installation techniques increase the speed of erection and often minimise the requirement for scaffolding. Unitised systems have higher direct costs and are less common than stick system. Nowadays the curtain walling to most prestige buildings is of this type.

Panellised systems

Panellised curtain walling comprises large prefabricated panels of bay width and storey height which are connected back to the primary

Materials and defects

structural columns or to the floor slabs. Panels may be of precast concrete or comprise a structural steel framework which can be used to support a variety of stone, metal and masonry cladding materials. The advantages of these systems are improved workmanship as a consequence of factory prefabrication, allowing improved control of quality and rapid installation with the minimum number of site sealed joints. Panellised systems are less common and more expensive than unitised construction. Panel systems often appear similar to unitised systems.

Variations

Structural sealant glazing is a form of glazing that can be applied to stick or unitised curtain walling systems. With structural sealant glazing, the double glazed units are attached to the grid framework with factory applied structural silicone sealant rather than by pressure plates and gaskets in a more traditional system. The attraction of this form of glazing is that it provides relatively smooth facades which are visually attractive.

Structural glazing typically comprises large thick single panes of toughened glass assembled with special bolts and brackets that are supported by a secondary steel structure. This form of glazing is often referred to as 'planar' glazing and is commonly used to form the enclosure to atriums and entrances.

Weather tightness

There are various methods of preventing rainwater ingress and these are discussed as follows.

Face sealed systems

Early curtain walling systems tended to be face sealed relying on a weatherproof outer seal to prevent water penetration. The seal must remain completely free of defects to prevent leakage paths occurring. Where there is no provision for drainage, any water that bypasses the outer seal could result in internal water ingress. With drained systems any water within the glazing rebate can then drain away within the framing system. There are also a small number of proprietary systems incorporating front zipper gaskets containing a large rubber gasket with a central press-in segment or zip which, when pressed into place, forces the gasket out onto the surface of the glass. This system does not normally have provision for water drainage.

Fully bedded systems

The systems are now largely obsolete and were utilised on the early forms of curtain walling. Fully bedded glazing is a face sealed system relying on the glazing rebate being completely filled with glazing compound to prevent the passage of water. They are therefore 'undrained'. Any voids within the bedding are a potential weak link for water ingress and early failure of the double glazed units.

Drained and ventilated systems

Most cladding designers now accept that it is difficult to exclude water and therefore provision for a small amount of leakage can be made within a drained system. Typically these dry glazed systems comprise an outer decorative cover plate, and an aluminium pressure plate with two narrow rubber oyster gaskets either side clamped against the glass or insulated infill panel. The pressure plates are screw fixed through a thermal break into the mullion or transom member. A further inner gasket between the glass and the mullion or transom provides a further seal.

In drained and ventilated systems the front gaskets provide an initial barrier. The rebates and cavities are drained and ventilated to the exterior to prevent the accumulation of any water that bypasses the outer seals. Drainage is usually via small holes or slots in the underside of transoms that drain water down through the mullions.

Materials and defects

Some systems also incorporate a foil faced butyl adhesive tape applied over the transom and mullion nosings directly beneath the pressure plate to serve as a secondary line of defence.

While these systems will accommodate a small quantity of water within the glazing rebates, it is important that the pressure plate is fixed to the correct torque so that the outer gasket seal forms a good seal against the glass.

Pressure equalised systems

Pressure equalised systems are an improved variation of drain and ventilated systems. Here the ventilation openings in the pressure plates are of an increased size to permit rapid equalisation of pressure in the glazing rebates with the external pressure thereby preventing water penetration of the outer face. Consider the following diagram:

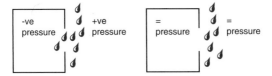

The rectangular box may be considered as the area around the glazing in a cladding system. If the pressure inside the box is less than the pressure outside, the water will be drawn in. If we can equalise the pressure in the box then the probability is that water will stay on the outside. In curtain walling systems we must provide a perfect seal around the inside of the window and must also provide the number of slots around the perimeter of the glass to enable pressure within the glazing rebate to equal that of the external air pressures almost instantaneously. Fundamentally therefore with pressure equalised systems the outer face is sealed as tight as possible against rainwater while the inner face is sealed as tight as possible against air inflow. Nearly all modern curtain walling systems are designed utilising the principles of pressure equalisation.

It is very difficult to identify the differences between a drained and ventilated system and a pressure equalised system. However for pressure equalisation to work properly, the various zones of pressure must not be too large. Thus it is common to consider the area around one glass pane as one zone and therefore drainage must be made from the transom members and not from the mullions. In practice it is very difficult to provide a fully effective pressure equalisation system and there is some doubt in the industry as to whether they are fully effective.

Double glazed units

Double glazed units are often referred to as 'insulating glass units' comprising two or more panes of glass spaced apart and hermetically factory sealed with dry air in the unit cavity. The air may then be flushed out and replaced with a range of other gases to improve thermal or acoustic performance.

The perimeter edge sealant prevents moisture from entering the unit cavity and holds the unit together. Two forms of edge seal configuration are single seal systems and dual seal systems. Single seal systems rely on the edge sealant to act both as the vapour barrier and as an adhesive bond to hold the panes of glass together. Dual seal glazing units rely on two seals, an inner seal to control water vapour transmission and an outer secondary seal to hold the glass tightly against the spacer bar. The combined properties of primary and secondary sealant provide high quality glass units. Nearly all units are now of this type.

The appropriate British Standard for Dual Seal Systems is BS5713.

Materials and defects

There is a wide variety of glass types that may be used.

Annealed glass is untreated glass manufactured from soda lime silicates. It is the least expensive and most readily available type of glass. Annealed glass breaks into sharp edged shards and is therefore considered to be unsafe in all fire and breakage situations.

Low-E coatings reduce the omission of long wave thermal radiation from the glazing and increase the reflection of this radiation. In the winter, solar radiation can be trapped within a room and reduce the need for heating. In the summer however heating can also occur and so low-E glass needs to be used in conjunction with adequate provision for ventilation. Most Low-E glass can be toughened and laminated.

Thermally toughened glass is formed by heating and then rapidly cooling or quenching annealed glass. Differential cooling and hardening across the thickness of the glass generates a compressive stress in the surface layer of the glass. Toughened glass is always a safety glass and compared with annealed glass is four to five times stronger in compression and bending. In failure, toughened glass shatters into small, relatively safe fragments. All annealed glass contains nickel sulphate impurities and as the glass is heated during the toughening process these impurities change state. Spontaneous breakage of the glass may follow. The failure of toughened glass in service can be reduced by heat soaking to encourage reversion of the impurities to the low temperature state before installation. Poor design and careless handling of glazing are much more common reasons for spontaneous breakage of toughened glass then nickel sulphite inclusions. As a minimum requirement, toughened or heat strengthened glass located at height or overhead should be heat soak tested.

Heat strengthened glass is formed by heating annealed glass and then cooling it under controlled conditions. Heat strengthened glass offers some of the strength of toughened glass but a reduced risk of failure due to nickel sulphite inclusions because of the reduced tensile stress in the glass. This glass is often also referred to as 'partially toughened'.

Laminated glass is formed by bonding together two or more panes of glass using a plastic interlayer. Any of the above forms of glass may be used in any combination. Upon failure, laminated annealed glass breaks into shards which are held together by the interlayer. Laminated glass may include one or more panes of toughened glass. If all panes are of toughened glass then the broken glazing will lose all structural integrity and may pull free from the pane unless properly secured. Laminated glass is recommended for the inner pane of overhead glazing and is considered to be safety glazing.

Surface finishes

The metal components of a curtain walling system will nearly always require finishes to provide protection against corrosion or for appearance.

To preserve the decorative and protective properties of any metal finishing, it is essential that atmospheric deposits are removed at frequent intervals, particularly those surfaces which are not exposed to the washing effects of the rain.

Quality of workmanship is particularly important and it is therefore essential to choose a reputable applicator preferably covered by a quality insurance scheme. Independent acceptance inspection testing can be undertaken to ensure compliance with the specification.

Organic coatings

Organic coatings are normally applied to either steel or aluminium and include polyester powder, PVDF, PVC, Plastol, and polyester. The most common organic finish for windows and curtain walling is polyester powder coating, however a range of wet applied finishes is widely used for opaque cladding panels. Polyester powder coatings may be applied to either galvanised steel or aluminium and are available in a wide range of colours. Polyester powder coatings are tough and abrasion resistant. Manufacturers often provide a guarantee for 15 to 20 years.

Anodising

Anodising is an electrolytic process that produces a dense, hard and durable oxide layer on the surface of aluminium. The oxide layer is porous and must be sealed to prevent staining but can be coloured by introducing dyes or chemical treatment before sealing. Anodised finishes are generally harder and more abrasion-resistant than organic coatings with an expected life of 50 years or more. However, anodised surfaces are susceptible to alkaline corrosion from contact with fresh concrete or mortar and rainwater run off from concrete surfaces.

Testing

Many standard and bespoke curtain walling systems are tested in laboratory conditions to determine the resistance to wind load, air tightness and water tightness. These tests are undertaken on a very small number of test panels assembled in factory conditions. By necessity they are a test of the design rather than on-site workmanship.

It is therefore critical that all installed curtain walling systems are subject to on-site testing to establish that the fabrication and installation has been undertaken to a satisfactory standard.

Water tightness on site is typically assessed using sprayed water from a hose in accordance with the AAMA Standard 501-94 or CWCT test methods. Ideally, the first areas to be tested should be among the first areas of each type of curtain wall to be constructed on site. Typically, the test areas are at least one structural bay wide and one storey in height, providing that all horizontal and structural joints or other conditions where leakage could occur are included.

Water is applied via a brass nozzle on the end of a hose that produces a solid cone of water droplets with a spread of 88°. The nozzle is provided with a control valve and a pressure gauge between the valve and nozzle. The water flow to the nozzle is adjusted to produce 22 +/- 2 litres per minute, producing a water pressure at the nozzle of 220 +/- 200 Kpa. Water is directed at the joint perpendicular to the face of the wall and moved slowly back and forth over the joint at a distance of 0.3m from it for a period of five minutes for each 1.5m of joint. There should be an observer of the inside of the wall, using a torch if necessary, to check for any leakage.

Mechanisms of water entry

Ways rainwater can penetrate

- ❖ kinetic energy
- ❖ surface tension
- ❖ gravity
- ❖ capillarity
- ❖ pressure differentials
- ❖ any combination of these.

Kinetic energy

This is the direct action of the wind carrying a droplet of rainwater with sufficient momentum to force it through a sealed joint. Prevention or drainage overcomes this. Prevention can be by baffles, by a durable seal or by a labyrinthine shape within the joint. Drainage collects the penetrating water and diverts it back to the outside.

Materials and defects

Surface tension

This can cause water to adhere to and move across surfaces. It is guarded against by drip edges or throatings along leading edges, and horizontal surfaces should slope down and out. Connecting components can also have appropriate grooves or ridges.

Gravity

This can take water through open joints that lead inwards and downwards. Reversing the slope overcomes this.

Capillarity

This occurs in fine joints between wettable surfaces. It is only severe when other mechanisms persist – for example, wind-assisted capillarity. In metal components it is resolved by capillary breaks within the joint surfaces.

Pressure differential

This frequently is the main mechanism. It is overcome by maximising the outer deterrent and minimising the pressure differentials. This is achieved by self-contained (compartmentalised) air spaces behind the outer skin, which are well ventilated to the outside.

Glazing – windows and doors satisfying the Building Regulations

The 2002 amendments to part L of the Building Regulations have brought about important changes in the specification of windows and doors. Essentially, compliance with the regulations will involve improving the insulation value of the frame (perhaps by reducing its size and increasing the extent of the better insulated glass panel) or by using more advanced glazing specifications such as low e coatings.

Specifically, measures could include the following:-

- ❖ Increase the width of the thermal break.
- ❖ Utilise an alternative type of thermal break.
- ❖ Provide alternative gasket design.
- ❖ Increase the air gap in the double glazed unit.
- ❖ Provide a gas filling to the glass.
- ❖ Provide an alternative spacer (warm edge technology).
- ❖ Utilise low e glass (either hard or soft coatings).
- ❖ Triple glazing.
- ❖ Dual double glazing.

Increase the width of the thermal break

The width of the thermal break will have a significant impact on the thermal performance of the frame. Pre April 2002 designs were often in the order of 12mm or less, but widths of 36mm are also realistic. Even with low e glass and an argon gas filling, a standard 4mm thermal break in a metal window will not meet the requirements.

Utilise an alternative type of thermal break

There are broadly two types of thermal break: those comprising a strip or 'web' of nylon 66 (polyamide) or glass fibre reinforced polypropelyne inserted into locating channels in the aluminium sections.

There is also those of resin which is poured into a single extrusion. When the resin has set the aluminium is then cut out.

By replacing a polyamide break with one made of polyurethane, one can reduce the u-value from say 3.95 W/m²K down to 3.72 W/m²K. However, polyurethane is not capable of accommodating the wider break designs.

Provide alternative gasket design

The thickness and positioning of gaskets, particularly those types, which also compartmentalise the glazing rebate, can influence thermal performance to a comparable degree.

Increase the air gap in the double glazed unit.

Many domestic DGU's are of 12mm width, with some timber window frames being only able to accommodate a 6mm air gap with two panes of 4mm glass. By increasing the width of the gap to 16mm, thermal performance is improved. Beyond 16mm there is little thermal benefit to be gained.

Provide a gas filling to the glass

Conventional DGU's are air filled, although the substitution of air with argon or krypton gas will lead to a marked improvement:-

- ❖ Air filled low e double glazed unit U=1.9 W/m²K.
- ❖ Argon filled low e double glazed unit U=1.6 W/m²K.
- ❖ Krypton filled low e double glazed unit U=1.35 W/m²K.

If the gas filled unit is well made it is possible that the gas will remain contained for many years. Estimates suggest that after manufacture, about 90% of the volume will be the specified gas with the remaining being air. A good unit should not loose more than a further 5% of gas over the next 25 years. However, there are no established test methods for establishing how much gas has been lost in service. The quality and effective life of gas filled units is therefore unknown at the present time.

Provide an alternative spacer (warm edge technology)

A high performance DGU such as those described above would not function as effectively as it could because standard aluminium spacer bars have a high thermal conductivity. By replacing the spacer bars with systems of low conductivity, it is thought that improvements of as much as 10% can be achieved. However, there are some concerns over the long term durability of these units. Additional types of warm edge spacers such as silicone foam cannot be tested under the methods described in prEN 1279-2 and this could restrict their use in the future.

Utilise low e glass (either hard or soft coatings)

Low e or low emissivity glass is made by coating one surface of a pane with a special metal coating. The treated pane is then encapsulated into a DGU, with the treated pane on the inner or room side. The coating reduces the amount of light that can pass through in the infra red spectrum, thus more heat is trapped on the inside of a room.

There are two types of low e glass:

- ❖ Pyrolitic hard coat; and
- ❖ Post applied soft coat.

Pyrolitic hard coat (for example, Pilkington k and SGG EKO Plus) is applied while the glass is being made. Thus the coating is very resilient, meaning

Materials and defects

that the glass can be toughened, cut and assembled without the need to remove the coating at the spacer bar contact point.

Soft coat (for example, SGG Planitherm) is applied post manufacture. Very high thermal properties will be dependent upon the type of coating used, but most soft coats cannot be toughened and are easily damaged. DGU's must be assembled within a short time of being cut. For these reasons, soft coatings tend to be more expensive, but some manufacturers will claim that they can provide improved thermal performance over hard coatings.

Triple glazing

A triple glazed window using float glass is not as effective as a DGU with Low e glass, and is much thicker resulting in heavier frame sections. However, a triple glazed unit with two low e panes can achieve 1.0 W/m≤K, while if the voids are filled with krypton, 0.7 W/m^{2}K can be achieved.

Dual double glazing

As its name implies, this involves setting two DGU's within a frame. However, the system would be very heavy with associated handling and cost implications. Thermal bridging at openings might be reduced because of the increased width of these systems.

Spontaneous glass fracturing

Nickel Sulphide is one of several chemical contaminants that can occur during the manufacture of glass. There is some debate as to its origin, but it is thought that it is due to the mix of nickel and sulphate impurities within the glass batch materials, the fuels or even the furnace equipment, and this creates polycrystalline spheres which vary from microscopic to 2mm in diameter.

All glass has some of these inclusions present; they are impossible to eliminate entirely and therefore they are not considered a product defect.

In untreated (annealed) glass they are not a problem. But when glass is heat treated (toughened or tempered), the inclusions are modified into a state which transforms with temperature and time and which is accompanied by an increase in volume.

In a majority of cases this has little effect but dependant on size and proximity to the centre of the pane where the forces are greatest, this can eventually cause the glass to break.

There is a theory that for an initial period of approximately one year after manufacture there are relatively few breakages. After this, the number increases for up to several years, thereafter decreasing in frequency. There have been reported incidences where fractures have occurred more than 20 years after the installation of glass.

Prior to pr EN 14179, a breakage rate of 1 per 5 tonnes of glass was thought an average level. However, this corresponds to only 200m^2 of 10mm glass. This, on a glazed roof of say 6000m^2, could be expected to have a failure rate of 30 breakages.

Panes in external situations are at greater risk. However, there have been a small number of cases where spontaneous breakage has occurred in internal glazing, remote from external influences, for example, panels to a staircase balustrade or an internal partition.

Although the safety risks are very small, because of the risk of falling debris, some companies will not recommend or supply and glaze toughened glass for sloped overhead applications where it will be used either as a single pane or as the inner leaf of a sealed unit.

Materials and defects

In 'heat strengthened' glass, Nickel Sulphide inclusion is not generally regarded as a source of fracture. The difference between this and toughened glass is the rate of cooling. In the former this is less rapid, reducing surface compressive strength and making it much less susceptible to the transformation of Nickel Sulphide inclusions. Offering 5.5 times the strength of annealed glass, in many circumstances it is a useful replacement for tempered glass. However, it is not a suitable substitute where safety glass is required.

'Heat soaking' is a quality controlled process which gives increased reassurance against the presence of critical Nickel Sulphide inclusions by subjecting the glass panels to accelerated elevated temperatures to stimulate the transformation of the crystals and thus initiate immediate failure. It is thought that this process identifies 90% or more glass, which might have subsequently failed after installation. The heat soaking process could be used either as a sampling method or as an additional treatment, which in the case of clear toughened glass could add up to 20% to the cost.

Heat soaking does not change any of the physical properties of toughened glass and therefore there is no means of distinguishing whether or not this process has been carried out. Current best practice dictates that specifiers should ensure that they specify 'heat soaked toughened glass to pr EN 14179. This standard reduces the anticipated failure rate to 1 per 400 tonnes of glass used.

Identification

When toughened glass is broken, the tensile stress is spread out from the source causing the pane to crack into small fragments (dicing). These fragments tend to be slightly wedge-shaped, emanating from the source of the fracture and are often held into position wedged against the frame due to their increased volume.

If the fracture is as a result of expansion of the Nickel Sulphide inclusion, those fragments immediately adjacent are more hexagonal and at the epicentre of the breakage the two larger particles form a distinctive butterfly shape linked by a central straight line crack. If large enough, the inclusion may be seen in the form of a black spec, or its presence may be confirmed by optical microscopy.

When carrying out an investigation, all possible causes of failure should be considered, including poor glazing tolerances and insufficient allowance for subsequent movement of the frame and any supporting structures. Possible causes may include deflection or rusting of steel frame, shrinkage of concrete frame, thermal movement, normal air pressures and even sonic booms. If the fracture is a result of impact or of local point loading, there should be evidence of local crushing.

The chances of installing a toughened glass pane, which may later fail due to the expansion of Nickel Sulphide inclusions, are very small.

Where it is essential, for reasons of accessibility or safety, that the pane should not fail, alternative forms of glass should be considered. One particular example would be in overhead situations – say in a shopping mall.

Testing

British Glass, (tel: 01142 686201) will test a failed sample to confirm the existence or otherwise of Nickel Sulphide inclusions. Samples should be sent intact with the epicentre protected with clear film.

Conservation

- Contaminated land — 212
- Radon — 215
- Energy conservation — 217
- Environment and specification — 218
- BREEAM — 220

Conservation

Contaminated land

Environmental due diligence is now commonplace in property transactions. The presence of contaminated land can adversely affect site value and rental income and can hinder transactions if not properly managed

With an increasing move away from greenfield development, it is no longer possible for the majority of investors or tenants to avoid owning or occupying some land affected by contamination, whether as a city centre property that has had a variety of past uses, or a new out of town development constructed on an old industrial site

Documentation is the key to maximising value and rental income and ensuring a smooth transaction. Documentation should be in place to demonstrate the following:-

- ❖ the site condition has been adequately assessed by an environmental assessment or review (commonly known as either an Environmental Audit or Phase I) and/or an intrusive site investigation (Phase II) and there are no significant information gaps;
- ❖ whether contamination is present (or is likely to be present), and the types of contaminants;
- ❖ where contamination has been identified, it does not represent a risk to the existing or proposed use of the site;
- ❖ contamination is not migrating off-site within groundwater;
- ❖ contamination does not represent a significant risk to groundwater and surface water resources or other sensitive receptors (eg sites of special scientific interest);
- ❖ contamination does not represent a risk of regulatory authority action (for example under Part IIA of the Environmental Protection Act 1990 - see below); and
- ❖ contamination does not represent a risk of third party action (for example, from adjoining land owners).

Environmental audit review

An environmental audit is based on background research and includes a site walkover. An environmental audit desktop study does not include a site walkover. A walkover would normally be advised to ensure all present day site issues are appropriately assessed. The research will normally include a review of:

- ❖ current site uses;
- ❖ historical site activities;
- ❖ environmental sensitivity;
- ❖ regulatory authority records; and
- ❖ a risk assessment and environmental risk rating.

Phase II ground investigation

'Phase II' means an assessment that is based on an intrusive physical investigation of the site, and usually includes chemical analysis and/or monitoring of soil, groundwater and surface water and land gas.

If an environmental audit or existing knowledge identifies a potentially significant contamination issue, then it may be necessary to conduct a Phase II intrusive investigation, to gain an understanding of whether contamination is actually present and whether it is likely to represent a significant risk

A Phase II investigation will normally include:

- ❖ a summary of the environmental audit findings;
- ❖ a description of the Phase II investigation methods;

Conservation

- data and observations recorded during the site work, including field evidence of contamination;
- data from chemical analysis of soil/water/gas samples; and
- interpretation of the results and an assessment of risk.

The Phase II investigation should be designed to address the specific issues raised by the environmental audit. The Phase II investigation should look for the range of contaminants highlighted by the environmental audit as possibly being present (for example, petrol on a petrol station, landfill gas on a landfill site etc).

The majority of sites undergoing development will require Phase II investigations, particularly where the previous use was industrial. Phase II contamination investigations can often be combined with geotechnical investigations for foundation design. Sites that are not being developed may also require Phase II investigations (for example, during transactions), depending on the specific circumstances.

Remediation

Remediation of sites can be achieved by removing sources of contamination, reducing levels of contamination or modifying the 'pathways' between the source and the identified sensitive receptors. The main approaches include:

- excavating and removing contaminated material off site to landfill (dig and dump);
- encapsulating or separating contaminated material on-site, by severing contaminant pathways with barriers (eg underground bentonite walls); and
- treating the contaminated material, either in-situ or after removal.

Remediation can often be combined with the redevelopment of the site, for example by excavation. Remediation strategies should be approved in advance by the local authority and the Environment Agency. Post-remediation validating sampling (eg of soil/groundwater) should be undertaken to document that the remediation has been effective.

Environmental insurance

Environmental insurance is increasingly being used in property transactions to cover risks associated with contaminated land. Environmental insurance can be obtained directly from specialist underwriters, or through an insurance broker. A broker will normally obtain quotations from a number of different underwriters, in order to negotiate the best insurance cover and premium for a client.

The most common type of environmental insurance covers regulatory and third party claims due to land contamination. Other types of insurance can provide protection against unpredictable costs should site remediation expenses prove difficult to quantify at the planning stage.

When taking out environmental insurance, one of the most important aspects is to understand what circumstances the policy will not cover.

Furthermore, as with many insurance policies, environmental insurance policies are often written in a language that is sometimes not at all clear.

Common exclusions from policies are:

- 'known' contamination;
- business disruption costs;
- remediation costs on change in use; and
- loss in value.

Key contaminated land legislation

The government believes that the planning process provides the best means of remediating sites, and in most cases the planning system is capable of handling such issues, for example by requiring Phase II site investigations or by requiring remediation to be approved by the local authority prior to development. A new contaminated land regime was introduced in 2001 in the UK (as Part IIA of the Environmental Protection Act 1990) intended to deal with problem contaminated sites that are not being developed and which would therefore not be dealt with under the planning system.

Definition of contaminated land under Part IIA

For the purposes of the Part IIA legislation 'contaminated land' means land where substances are present in, on or under, the land that are causing, or are likely to cause, significant harm or pollution of controlled waters. Many sites that are contaminated will not fall within the definition and will not be classified as 'contaminated land' under Part IIA. However, such contamination could still have implications for owners and occupiers (for example, in terms of affecting the saleability and marketability of a site) and may still require a contaminated land assessment.

Who is liable under Part IIA?

- ❖ Remediation notices are served in the first instance on 'Class A' persons, ie polluters or knowing permitters of contamination. If these responsible parties cannot be found, the current owner or occupier may be responsible (although for a more limited range of liabilities) – 'Class B' persons, Several parties may be implicated.
- ❖ Sellers can avoid liability where there are payments for remediation, with an explicit statement in the sale contract that a purchaser is being paid to clean up land or that the purchase price is being reduced to reflect the contaminated state of the land.
- ❖ Sellers/landlords can also avoid liability by selling with information – giving the purchaser (or tenant under a long lease) the necessary information to identify contamination before buying. Where transactions have occurred since 1990 between large commercial organisations, the granting of permission by the seller for the buyer to carry out its own investigations as to the condition of the land, is normally sufficient to indicate that the buyer had the necessary information.

How the Part IIA regime works

The Part IIA regime is enforced by local authorities, or by the Environment Agency (EA) /Scottish Environment Protection Agency (SEPA) in the case of 'special sites'. Special sites are the most seriously contaminated sites which include those sites affecting water supplies or major aquifers, and include military land, oil refineries and nuclear sites.

Once land is identified as being contaminated, local authorities (or EA/SEPA) must achieve clean-up. This can occur through the planning process, via voluntary remediation by the responsible party (see definitions above), or if necessary by serving remediation notices. There is a three month consultation period before a remediation notice is served, to encourage voluntary remediation (except in emergencies). Local authorities and the government are encouraging the use of informal approaches to the management of contaminated land, such as Remediation Statements (voluntary agreements between responsible parties and the regulator to carry out remediation). This may avoid the need for the sites to be entered onto public registers of contaminated land.

Conservation

Progress with Part IIA implementation

Although implementation of the regime has been slow, 109 of the 119 English Local Authorities now have contaminated land inspection strategies. However, only 43 of these are on schedule to meet their target dates (according to the National Society for Clean Air - NSCA). The latest available government figures (published in June 2003) indicate that of the estimated 50,000 to 300,000 contaminated sites UK-wide, English local authorities have formally designated only 58 sites as 'contaminated' under the Part IIA definition. A further 14 sites have been designated as 'special sites'.

A significant amount of remediation has been carried out by the regulators (at least 63 sites in 2002-2003). However, the number of remediation notices served by local authorities remains very low. In most cases, remediation is undertaken by site owners via the planning system with infrequent use of remediation statements. A survey published in July 2003 by NSCA reported that none of the local authorities responding had issued any remediation notices at all.

Radon

Radon is a radioactive gas that occurs naturally in the earth. It has no taste, smell or colour and detectors have to be used to test its presence. When concentration levels rise within buildings, it can pose a serious risk to health.

Radon occurs when uranium decays and becomes radium. When radium decays, it becomes radon. Uranium is found in small quantities in all soil and rocks. It can also be found in building materials derived from rocks.

Radon rises from the soil into the air. Outdoors, radon is diluted and the risk it poses is negligible. Problems occur when it enters enclosed spaces, such as buildings, where concentration levels can build up.

Radon is everywhere but usually in insignificant, variable quantities. There are some areas in the UK where geographical effects result in higher levels.

The National Radiological Protection Board (NRPB) has produced maps of radon affected areas, which include Derbyshire, Devon, Cornwall, Northamptonshire, Somerset, Yorkshire and the Lake District in England, Grampian and Highland regions in Scotland and County Down and Armagh in Northern Ireland, and Bristol, Newport and West Cardiff in Wales.

The NRPB has set threshold levels for both commercial and residential properties. Detectors should be installed to ascertain the level of radon present.

If the readings exceed the threshold level, remedial works will be required in order to reduce the effects of radon. They offer a search service and report, which specifies whether a property is in a radon affected area and the probability that radon will exceed the action level. However just because a property lies within one of these areas it does not necessarily mean that it will have a problem with radon. The British Geological survey also states whether radon is likely within a particular area as part of its Address Linked Geological Inventory.

Identification of radon

Radon levels vary appreciably with time, so prolonged measurements are required for reliable results. Short measurements can be misleading, low or alarmingly high. The government recommends that people in affected areas test their property for a period of three months using passive monitors in order to provide a reliable estimate of the average radon level. Passive monitors are easy to use, inexpensive and available from DEFRA. A good rule of thumb is one detector per 100m^2 of floor area. In larger open

Conservation

planned work areas such as a production area in a factory or an open plan office, the number of detectors may be reduced to one per 500m² of floor area.

Remedial action for high radon levels can be quite straightforward. The best approach is to prevent radon entering the building from the ground by altering the balance of pressure between the inside and outside.

This can be achieved by carrying out the following:-

- ❖ Install a small sump pump below the floor and connect to a low power fan in order to extract the air and reduce the pressure under the floor. This is known as an active sump (a passive sump relies on natural forces to drive air through the system). To minimise inconvenience, the sump may be outside the building with a pipe through the wall. A sump is a reliable and effective remedy that can reduce radon levels by at least a factor of ten. Multiple sumps can be used in large buildings.
- ❖ Improved ventilation under suspended timber and concrete floors. New airbricks are installed, sometimes together with a fan. This system again limits the amount of radon entering the property.
- ❖ Increase the pressure in the building by blowing air (called positive pressurisation) from the roof space with a small fan. Best results are in buildings with low natural ventilation. This is a reliable remedy that can at least halve radon levels. Secondary benefits may include a reduction of other indoor pollutants such as carbon dioxide, reduced condensation and a 'fresher' indoor environment.
- ❖ Alternatively, one may seal ducts, joints and cracks in the floors although this is rarely effective by itself and always laborious. It is helpful to close large openings when a sump is used.
- ❖ Ventilation (other than positive ventilation).
- ❖ Install a membrane barrier. This is very difficult to successfully achieve in an existing building.

For domestic properties the DETR recommend six main ways to reduce indoor radon levels to significantly below the action level of 200 Bq/m³.

Remember that it is the average exposure to radon that matters. Short exposure at high levels is not important if over the long term your average exposure is low. This means that you should have time to plan for the solution that is best for the client, property and the radon level. But having found the best solution, it should be implemented as soon as practical.

It is best to stop radon entering a house or, if that is not possible, to try and remove it if it gets in. Solutions are similar to those recommended for commercial properties, but with subtle differences:-

- ❖ Install a radon sump pump. The system limits the amount of radon that enters the house and for a typical house it is by far the most effective method. Modern sumps are often constructed from the side of the house so there is no disruption inside.
- ❖ Improve ventilation under suspended timber floors.
- ❖ Use positive ventilation. This system is designed to change the air pressure in the house by blowing air in from the loft level. The system both dilutes the radon to acceptable levels and stops some of it getting in.
- ❖ Seal cracks and gaps in the floors.
- ❖ Change the way the house is ventilated. This solution is only suitable in quite special cases and has drawbacks.
- ❖ Install a membrane barrier.

Conservation

When action has been taken and radon levels reduced, it is recommended to arrange routine checks on fans and other equipment. Periodic measurements of radon should also be made for overall assurance that levels remain low. These guidelines apply to all types of buildings.

Radon and the Building Regulations

With the new understanding of radon risk, the government legislated that houses built since 1988 in parts of Devon and Cornwall and 1992 in parts of Somerset, Derbyshire and Northamptonshire had to have radon protection measures built in.

Two zones of risk were allowed for. First the primary zone (the area with the highest risk). Requirement C2 of Schedule 1 of the Building Regulations requires that each house has a radon proof area, together with other precautionary measures that can be upgraded if a risk shows high radon levels. In the secondary zone (where the risk is lower) only precautionary measures must be built in. If a house has precautionary measures, upgrading them (for example, adding a fan to a sump and pipe system) could solve the radon problem quickly and simply.

The BRE has published guidance on protective measures for new dwellings in support of the Building Regulations entitled BRE Report 'Radon: Guidance on Protective Measures for New Dwellings'. No equivalent advice has been prepared for commercial buildings, but it is recommended that similar measures will be applicable for non-domestic buildings.

Energy conservation

Some of the ways in which energy efficiency can be enhanced are outlined below:-

- **Greater thermal insulation.** This is one of the most cost-effective ways of increasing a building's energy efficiency. With new buildings, obviously the starting point is to meet the standards of thermal insulation set out in the current building regulations. However, there is a strong argument for further increasing thermal insulation and the level chosen will depend upon a variety of factors including the pay-back period required, the user's pattern of occupation and whether environmental concern outweighs strictly financial considerations. Our existing buildings offer great scope to increase insulation and this can be carried out during refurbishment or maintenance periods when better use can be made of access equipment and other site overheads.
- **Efficient lighting.** In commercial buildings, lighting costs usually run between 40% and 50% of total energy costs. Highly efficient lighting that can significantly reduce running costs is now available for both new and existing installations.
- **Efficient services.** These require preventative maintenance to ensure that all plant is operating at maximum efficiency. When plant requires renewal, consideration should be given to alternatives, such as condensing boilers or combined heat and power plant, or even whether the plant is required at all; many are re-evaluating the need for air conditioning.
- **Building management systems.** These monitor and control all service installations. Due to recent technological advances, these systems are becoming less expensive and can therefore be installed on smaller properties. Even where a full BMS is inappropriate, simple systems are available which control single services such as lighting management systems.
- **Using locally sourced materials.** This minimises energy consumption when transporting materials to site.

- **Building materials.** Those with a long life expectancy imply energy efficiency because they make good use of resources. The manufacture of building materials entails energy consumption and this varies widely depending on the product. Unfortunately, there is insufficient research into the energy consumed during the manufacture of building materials to allow choices to be made with a great deal of confidence. In the meantime, a useful rule-of-thumb is that the greater the degree of processing or manufacture, the greater the energy consumed.
- **Making use of solar gain.** Even with their humblest dwellings, our ancestors frequently designed their buildings to make use of solar gain. Generally this entails the use of large areas of glazing on southern elevations and minimal openings to the north. Increasing numbers of building designers are re-interpreting these techniques.
- **The use of soft landscaping.** Trees and other planting can conserve energy in buildings by minimising heat gains and losses. A screen of deciduous planting at the south of a building will filter strong summer sun but will allow for natural heat gain from weaker winter sun. Soft landscaping also has an important part to play in influencing the microclimate around buildings by reducing wind speeds.
- **Out-of-town schemes.** Many of these schemes are built with high levels of thermal insulation and efficient building services. Nevertheless, because of the fuel used in transporting building users from their homes, the whole scheme may be very inefficient in energy terms.

Increased energy efficiency is available at little extra cost – it is just a matter of adopting an environmental train of thought. Increased awareness of these issues by players in the property market and pressure of legislation should be seen as an opportunity to create more energy efficient buildings.

Environment and specification

The environment does not lend itself to simple right-or-wrong selection criteria of materials. However, careful specification does play a part in a complete design and operating philosophy for environmentally responsible building. There are, of course, environmental aspects to the use of all materials but for reasons of space only two model specification clauses are included here; timber and chlorofluorocarbons.

Timber

All references to timber contained within the specification are to be obtained exclusively from sustainable sources. The contractor is to provide evidence, by a supplier's certificate and labelling for each consignment delivered to site, that the timber is from such a source.

The certificate and label should include the following information:

- the species and country of origin;
- the name of the concession or plantation;
- a copy of the forestry policy; and
- shipping documents confirming the source.

The Good Wood Seal of Approval by Friends of the Earth would suffice. Information on appropriate suppliers can be obtained from the Timber Trades Federation, Friends of the Earth and the International Timber Trades Organisation.

Conservation

CFCs, HCFCs and halons

In line with current good practice, the following clauses relating to chlorofluorocarbons (CFCs) hydrochlorofluorocarbons (HCFCs) and halons, their removal and restrictions on use, shall be deemed to have been allowed for in any tender or estimate offered.

Existing plant

Where CFCs or HCFCs are identified as being contained within existing air conditioning or refrigeration plant, the contract administrator is to be informed immediately and instructions obtained.

Where the plant is scheduled, or instructed subsequently, for removal then under no circumstances is the gas to be dumped by venting into the atmosphere. The gas is to be collected for recovery/destruction by a specialist firm.

The contract administrator is to be informed in writing of the specialist undertaking the works, the date for removal and be provided with a copy of the recovery/destruction certificate.

New plant

The contractor shall receive and transmit to the contract administrator documentary evidence from suppliers, sub-contractors and designers of all new installations that no new or re-used plant contains refrigerants with an ozone depletion potential of more than 0.06 (or with any ozone depleting potential). Furthermore, compounds with the lowest possible ozone depleting potential are to be selected where there is a choice.

As an alternative, consideration can be given to the use of absorption chillers or ammonia chillers.

Halon fire fighting systems

The general requirements for de-commissioning the systems shall be the same as for CFCs and HCFCs in plant.

If the system is to be tested, then compressed air or some other non-ozone depleting gas shall be used.

If the system is to be recharged, a leak detection system should be installed. However, the preference is for some other form of fire extinguishing system wherever possible. Depending on the circumstances options include inert gases (Inergen, Argonite, etc), carbon dioxide and water fog/mist systems.

Hand-held fire extinguishers

All fire extinguishers on site supplied by the contractor, or specified to be supplied in the Schedule of Works section shall be either powder, foam, carbon dioxide, water spray or some other type without the use of Halon.

Additional measures

CFCs and HCFCs are used in the manufacture of a variety of other products including insulating materials, carpets, furnishings and aerosol sprays. They must not be used unless specifically instructed by the contract administrator.

Conservation

BREEAM

The Building Research Environmental Assessment Method (BREEAM) is the world's most widely used system for assessing, reviewing and improving a range of environmental impacts associated with buildings.

Since its launch in 1990 BREEAM has been increasingly accepted in the UK construction and property sectors as offering best practice in environmental design and management. Buildings are assessed against performance criteria set by the BRE and awarded 'credits' based on their level of performance.

The building's performance is then rated as pass, good, very good or excellent.

BREEAM covers a range of building types: offices, homes (known as EcoHomes), industrial units and retail. Other building categories can be assessed using a bespoke version of BREEAM.

BREEAM 2002 was launched on 31 August 2002 by the Building Research Establishment and all new assessments are carried out under this scheme.

BREEAM for Offices is updated every year to ensure best practice and relevance to changing standards and regulations. The challenge to achieve the highest rating has increased as targets are being continually raised.

Developers and designers can utilise BREEAM for a range of reasons including creation of better environments for people to work in, increased building efficiency, improved marketability, increased value and rentals and also as a checklist for comparing buildings.

Clients and developers can use BREEAM as a tool to define and specify the environmental and sustainability performance requirements of their buildings at the briefing stage. Agents can use the rating to promote the environmental credentials and benefits of a building to potential owners and tenants. Designers can use BREEAM as a method to improve the performance, environmental and sustainability aspects of buildings.

A BREEAM Office assessment of the building fabric and services is undertaken plus, as appropriate, the quality of the design and procurement and also management and operating procedures.

The scheme requires a commitment to a number of areas, listed below, that are reviewed by independent assessors who are trained and licensed by the BRE.

- Management: overall policy, site management via the Construction Confederation Considerate Constructors Scheme and procedural issues.
- Health and well being: both internal and external issues affecting occupants health.
- Energy efficiency including operational energy and carbon dioxide issues.
- Transport: carbon dioxide and location related factors.
- Water consumption and efficiency.
- Materials: environmental implications and life cycle impact.
- Land use regarding greenfield and brownfield sites.
- Ecology including enhancement of the site as well as ecological value conservation.
- Pollution of air and water.

To achieve an excellent rating the most cost effective way is to address the main issues at the earliest point of the design process with input from the full project team.

Conservation

A BREEAM assessor can be used to coordinate and collate input from the team and to track the development of ideas. The assessor can also give advice about BREEAM to the entire project team at the start of the project.

As the scheme progresses, the assessor can provide specialist advice on the specification of products to achieve particular BREEAM credits, undertake preliminary BREEAM assessments to assess the predicted rating and provide a sustainability report inclusion with submission for planning approval.

At completion a certificate is awarded and this can be used for promotional purposes.

Maintenance management

- Primary objectives of maintenance management   224
- A systematic approach to maintenance management   224
- Condition surveys   227
- 'Best value' in local authorities   229
- Sources of information in maintenance management   230

Maintenance management

Primary objectives of maintenance management

It is essential that property owners and managers allocate sufficient time and resources to maintenance. Managers need to be aware of the full extent of maintenance liabilities and how much money should be spent and when. Maintenance management requires a systematic approach to ensure high standards, value for money and management control.

Buildings comprise a number of elements. Their constituent materials and components have a range of life expectancies that in most cases will be shorter than the life of the building as a whole. Maintenance is therefore inevitable and arises from failure at this component level.

The primary objectives of maintenance are:-

Protection of the investment to ensure high utilisation of the building and its long life

Maintenance affects the profitability of a commercial organisation in a number of ways. First there are the direct costs of labour, plant, materials, and management. The second category is indirect where inadequate maintenance of buildings prevents the organisation from functioning properly.

Safeguarding the return on the investment

Poorly maintained commercial buildings fare unfavourably in the market when compared with their well maintained equivalents. The run down of an investment from lack of maintenance will discourage tenants from wishing to remain in occupation. This will have an effect on rent levels and encourage assignments and vacations on termination of leases.

The control of costs

Timely, planned maintenance enables efficient use of resources.

Establishing a safe working environment

The safety of the building and establishing a safe working environment must be the first priority of the maintenance manager.

Maintenance closes the gap between the actual state of the building and the acceptable standard. The acceptable standard will be dependent upon a number of factors, which may include:

- ❖ statutory requirements (health and safety);
- ❖ tenant or occupant satisfaction;
- ❖ minimising loss of production;
- ❖ morale of users, employers and customers; and
- ❖ public image.

A systematic approach to maintenance management

A systematic approach to maintenance management has a number of elements.

Policy

A building may be an asset to an investor or a resource to the user. The maintenance policy defines the objectives that maintenance of the

Maintenance management

building sets out to achieve. It is a dynamic concept, subject to change just as the plans and objectives of the user or organisation will change.

Standards have to be defined. The use of 'normal standard' is inadequate. The acceptable range of performance of any element of a building will depend upon the relationship between each of its functional requirements and the use of the building as a whole.

The policy must therefore contain objective criteria to define what constitutes failure or non conformance in each category. The policy must identify those activities that are sensitive to the physical condition of the building and those building elements that play a significant role in providing the necessary conditions.

Once these components are identified, the acceptable delay time in correcting any failure can be assessed.

The statement of policy constitutes the brief for the maintenance manager. It should cover future requirements of the buildings, changes of use, statutory and legal conditions, maintenance cycles, required standards and acceptable response times for breakdown or failure.

The policy must be regularly reviewed and amended as necessary. Consideration of a maintenance policy often shows that there may be some conflict between parties with different interests, for example, between landlord and tenant.

Survey and data gathering

In order to measure how the buildings compare against the policy, and how to close the gap, it is essential to review all existing maintenance information and to undertake regular condition surveys. These identify the condition of the asset and record the status of the building at any one time. A condition survey has the specific purpose of:

- identifying maintenance needs;
- recording the priority of the need;
- recording proposed remedies and quantities of items requiring attention before the next survey; and
- predicting the scale of items requiring attention after the next survey.

Planning of work

Having agreed a policy for the organisation and identified work to be done, the first function of the maintenance manager is to formulate a maintenance plan. The objective of planning is to ensure that work is carried out with maximum economy. A key function of maintenance management is to achieve positive control over the work and to avoid overloading or inefficient utilisation of resources. Planned maintenance consists of preventive and corrective work.

- **Preventive maintenance**

    That which is carried out at predetermined intervals and is intended to reduce the probability of failure.

- **Corrective maintenance**

    That which is carried out after failure has occurred and is intended to restore an item to a state in which it can perform its required function.

Generally speaking, the most economic plan in direct cost terms would be the one that maximises preventive maintenance. However, by definition, this form of maintenance takes place before failure has occurred and some degree of useable life has been wasted. This wasted life has a value that can be costed and must be added to the direct cost of the maintenance

Maintenance management

resource used. The split of work between preventive and corrective maintenance is a management decision. The actual proportion of each will be determined by reference to the maintenance policy.

Plans must take into account practical considerations and the availability of resources. They are prepared with different time horizons for different purposes:-

- ❖ Long term – for strategy, to establish general expenditure levels and to profile the maintenance demands of a property or portfolio over an extended time horizon.
- ❖ Medium term – addressing demands that are predicted within the next five years and to refine budgeting and the assessment of workload.
- ❖ Annual – for work and resource allocations.

Organising

The construction of an organisation and a control system capable of ensuring the implementation of the plan.

The main constituents of a maintenance organisation:-

- ❖ Resources – ie labour, materials, plant and management. These may be either directly employed or obtained through contracts with outside contractors.
- ❖ Administration – a staff structure for coordinating and directing resources.
- ❖ Work planning and control – the system for work, budget, cost and condition control. At its heart lies a documentation procedure, an information base to support effective decision making.

  The documentation system will include a number of elements such as an asset register, a property information base, and preventive maintenance documentation together with a system for initiating and controlling works: works orders, work request forms, etc. These systems are generally computerised. There are a number of proprietary computer programs available. Such programs revolve around relational databases that can be tailored for strategic work, planning and budgeting right through to processing of day-to-day orders for reactive maintenance.

Procuring

A contract and procurement strategy must be established. The strategy will be driven by the nature of anticipated and known workloads and the need to achieve maximum efficiency of operation.

The workload may include the need to provide reactive maintenance cover for unforeseen or accidental repairs, 24 hour emergency call out cover, planned maintenance contracts and small items of maintenance work packaged into cost-effective contracts.

The decision about which contract and procurement strategy to adopt will depend on a number of factors including the geographic spread of the portfolio in question; the preferred type of model arrangements to be implemented; and the number and type of resources available to the maintenance manager.

Monitoring

Monitoring and auditing is essential to ensure that the maintenance management system is functioning properly and achieving the requirements of the policy statement and if not, to record deficiencies and initiate corrective action.

Maintenance management

The maintenance manager needs to monitor the system and organisation to ensure that quality and value for money is being achieved. The audit is a post examination of maintenance work and procedures, not dissimilar to a financial audit. The audit can be condition, technical, systems or design based.

The condition audit involves the checking of planned maintenance schedules against the actual condition of the building.

The aim of the technical audit is to examine a sample of maintenance activities to investigate how each task has been approached and dealt with.

The systems audit will analyse maintenance management practices. It will establish the true nature and extent of the database being used and how this is stored and accessed. It will look at overall policy and whether or not there are realistic plans and programmes in existence, at the method used for budget preparation, budget control, and for feedback of information to assist managers in future decision-making.

The design audit concentrates on the interaction between design and building performance. In most cases it is a question of lessons to be learned for the future.

Condition surveys

Property assets are extremely important to owners, investors and occupiers and therefore need to be properly maintained and managed (see also 'maintenance management'). Research indicates that UK business fails to understand the importance of it's property assets, an important aspect of which is building condition which deteriorates with time, reducing value and utility.

Condition surveys represent an essential means of collecting data and ultimately reporting on the physical condition of a property or portfolio of properties, as well as addressing performance and statutory compliance issues if required.

The RICS Guidance Note (out of print) defines a condition survey as follows:

> The collection of data about the condition of the building, part of a building, estate or portfolio assessing how that condition compares to a pre-determined standard, to identify any actions necessary to achieve that standard now; and maintain it there over a specified time horizon; the purpose being to support management decision making.

Condition surveys are of benefit to everyone involved in the ownership, occupation and investment in property. For owners, occupiers and managers, these surveys provide a database of information identifying maintenance and repair work as well as the time scale for undertaking that work allowing budgeting, assessment of compliance and strategic planning. For investors, it represents an audit to measure the condition of their asset, while for purchasers it assists in asset valuation by identifying repair and dilapidation liabilities. At strategic/policy making level, it will enable both a view of policy effectiveness, and allow implications of regulatory change to be established.

Format and content

The agreed brief should define the number of properties to be inspected; for example, for portfolios of large numbers of similar types of property, a sample is often taken rather than surveying the full portfolio. It should confirm whether the surveys will include the internal and external building fabric, building services, infrastructure items such as roads and main services; hard and soft landscaping and boundaries; and the extent and detail of statutory compliance items. Agreement should be reached on the

Maintenance management

depth of the survey and the approach to complex defects which will often need to be noted as requiring further investigation, as well as whether any specialist tests and surveys form part of the requirement. The method of costing items should also be established.

The format and content of the reports will inevitably vary across clients and will relate to how they will aim to use and present the data, how they wish to prioritise the works identified, and what type of timescale the condition surveys are considering. Typically, condition surveys tend to consider the condition and works required over five to ten years but may address much longer periods.

The format and content of the surveys will also depend on how the data is to be collected and stored. With the availability of powerful computerised databases, and hand-held computer technology for gathering data, clients may not in fact require much in the way of traditionally written reports, while others would prefer to have a textual report with a basic chart or spreadsheet indicating building assets, building description, condition, priority of remedial works, along with a time frame and budget cost.

It is recommended that there should always be a written report summarising the findings of the survey in addition to confirming the limitations and the brief. It should include a summary of the results, possibly with easily understood graphics, as well as suitable cross reference to electronic files that may have been provided in the process.

Priorities

The classification of repair priorities will also vary but, typically, items affecting health and safety will have the highest priority, while works necessary to maintain an element in repair, or those maintaining civic/corporate standards may be less important. For example:-

- ❖ **Priority 1/serious/unavoidable** – Typically relating to health and safety compliance; items requiring immediate attention to avoid a major breakdown, serious hazard or critical deterioration leading to possible closure of the building.

- ❖ **Priority 2/poor/essential work** – Typically works that if neglected might lead to damage to the value of the property, and premature replacement.

- ❖ **Priority 3/fair/necessary work** – Where neglect might affect rental income, the condition of the element is less than adequate and defects need to be remedied within two/three years.

- ❖ **Priority 4/adequate/desirable** – Works necessary to maintain an element in repair but probably not in the immediate short term. Might include items that would improve working conditions or generate financial benefits such as energy conservation.

In order to avoid ambiguous interpretation of repair items, it is wise to keep the number of categories to a minimum. Whatever categories are adopted should be clearly defined with the survey team appropriately briefed and trained.

The subsequent use of data collected depends very much on how it is stored. It is essential to use spreadsheets or preferably databases that allow data to be filtered and sorted with enquiries run as appropriate to the beneficiaries of the surveys. Prioritisation of condition should, preferably, be undertaken using computers as this enables items to be evaluated with more than one factor, typically not only physical condition but also consequential effect on the asset and the building occupant, as well as management factors and policies.

Maintenance management

Frequency of surveys

It is unfortunate that often expensive condition survey exercises are undertaken providing what effectively results in a 'snapshot' of the condition at the time of the survey but thereafter the data is not used or updated. In the case of a property transaction, this 'snapshot' assessment of repair liability is appropriate but to ensure cost effective long term maintenance of building assets, the condition data should be periodically updated. British Standard 8210: 1986 suggests in-depth surveys on a five-year programme, supplemented by two-yearly inspections on a more superficial basis. Clients may choose a different profile but generally the view is that, for maintenance purposes, intervals of a maximum of five years are acceptable. Whether to survey entire portfolios in one go, or to perhaps undertake surveys on a rolling programme on a five-year basis, will depend on many factors, not least the funds available to commission the survey in the first place. Arguably, mass condition surveys of entire portfolios required in a short period may suffer in terms of quality, but if carried out correctly will give an earlier fuller picture than a rolling programme.

Further reading

Royal Institution of Chartered Surveyors Guidance Note - Stock Condition Surveys

British Standard 8210 (1986) British Standard Guide to Building Maintenance management.

'Best value' in local authorities

What is 'best value'?

The government has defined best value as "a duty to deliver services to clear standards – covering both cost and quality – by the most economic, efficient and effective means available". This clearly has a direct bearing on construction related services provided by local authorities

The government will require local authorities to publish an annual best value performance plan (BVPP) covering the entire range of the authorities' services. This is intended to be a public document and will include an assessment of the authorities' past and current performance against nationally and locally defined standards.

The performance plan will be the main instrument by which local authorities will be held accountable to the local community for delivering best value.

How are best value studies carried out?

The Audit Commission appoint an external auditor to undertake an audit of the authorities' BVPP. In the case of construction related services this is likely to be a professional consultant in the relevant field. The audit will cover:

- ❖ compliance with legislation and guidance;
- ❖ performance information; and
- ❖ continuous improvement strategies (the 4 Cs - see below).

'The 4 Cs'

Authorities are required to carry out a Best Value Review as part of the BVPP this includes putting in place strategy for continuous improvement implementing the '4 Cs':

Maintenance management

- ❖ **Challenge** – why and how the service is being provided.
- ❖ **Compare** – with others' performance (including organisations in the private sector) across a range of relevant indicators.
- ❖ **Consult** – with local taxpayers and service users with the view to setting new performance targets.
- ❖ **Compete** – as a means of securing efficient and effective services.

(More detail regarding the methods of inspection can be seen on the Audit Commission website – www.audit-commission.gov.uk or www.local-regions.detr.gov.uk)

Continuous improvement for construction and property services

A Service Improvement Plan (SIP) is one which identifies areas of potential improvement from:

- ❖ those services already subjected to a Best Value Review;
- ❖ advice from the Audit Commission inspection service; or
- ❖ advice from an external consultant by way of a best value 'healthcheck' (commonly termed a 'critical friend').

Depending on the particular service, the following core issues will be included in the SIP:

- ❖ Description of action – to be taken to improve service.
- ❖ Target – either qualitative or quantitative.
- ❖ Target and timescales – timescale for achieving the improvement(s).
- ❖ Likely effect – impact of the improvement identified.

Other issues which are likely to be included in construction related services could be:

- ❖ current performance;
- ❖ national benchmarks;
- ❖ resource requirements; and
- ❖ monitoring and review strategies.

This overview of best value is intended to give some insight into government efforts in improving services to the taxpayer. When considering the detail of Service Improvement Plans for construction and property related services it is recommended that consultancy advice be sought.

Sources of information in maintenance management

BMI/BCIS

Building Maintenance Price Book, published annually
SR 328 The Economic Significance of Maintenance. March 2004
SR 331 Review of Maintenance Costs 2004
SR 332 Review of Occupancy Costs 2004

BSRIA

Application Guide 1/87.1 Operating and Maintenance Manuals for Building Services Installations 1990

Maintenance management

Application Guide 4/89.2 Maintenance Contracts for Building Engineering Services second edition 1992

Application Guide 24/97 Operation and Maintenance Audits 1997

Application Guide 1/98 Maintenance Programme Set-up 1998

Application Guide 20/99 Cost Benchmarks for the Installation of Building Services Parts 1-3 1999

Application Guide 4/2000 Condition Survey of Building Services 2000

Guidance Note 2/2004 Computer-based Operating and Maintenance Manuals - Options and Procurement Guidance 2004

NJCC

Procedure Note 16 Record Drawings and Operating and Maintenance Instructions and the health and safety file. Second Edition June 1996 RIBA Publications

Note: NJCC no longer in existence, but guidance notes still in general use.

RICS

Building Maintenance: Strategy, Planning and Procurement – A Guidance Note, 2000, RICS

Computerised Maintenance management Systems – A Survey of Performance Requirements, May 1999, RICS Research paper

Miscellaneous

Building Maintenance management, Chanter & Swallow, second Edition August 2000, Blackwell Science

Building Maintenance and Preservation, Mills 1996, Butterworth-Heinemann Ltd.

Lee's Building Maintenance management, P Wordsworth, fourth Edition November 2000, Blackwell Science

Maintenance Management – Its Auditing and Benchmarking – A Kelly – Oct 2002

Maintenance management

British Standards in maintenance management

BS 3811	Glossary of Terms in Terotechnology	1993
BS 3843	Guide to Terotechnology	1992
	Pt2 Introduction to techniques and applications. Pt 3 Guide to the available techniques	1992
BS 6150	Code of Practice for Painting Buildings	1991
BS 6270	Code of Practice for Cleaning and Surface Repair of Buildings Pt 3 metals (cleaning)	1991
BS 7543	Guide to Durability of Buildings and Building Elements Products and Components	2003
BS ISO 15686	Buildings and constructed assets. Service life planning. General principles	2000
	Pt2 Service Life Prediction Procedures	2001
	Pt 3 Performance Audits and Reviews	2002
	Pt 6 Procedures for considering environmental impacts 2004	2004
BS 8210	Guide to Building Maintenance management	1986
BS 8221	Code of Practice for Cleaning and Surface Repair of Buildings	
	Pt 1 Cleaning of natural stones, brick, terracotta and concrete	2000
	Pt 2 Surface repair of natural stones, brick and terracotta	2000

Cost management

➡	Urban regeneration	234
➡	VAT in the construction industry – zero rating	235
➡	Capital allowances for taxation purposes	239
➡	Land remediation allowances	242
➡	Tender prices and building cost indices	243
➡	Rating revaluation 2005	244
➡	Cost management	245
➡	Life cycle costings	246
➡	Reinstatement valuations for insurance purposes	252

Cost management

Urban regeneration

What is urban regeneration?

As cities and urban conurbations grow they constantly change to respond to demand and supply according to prevailing economic, social and environmental factors. Many thriving communities lose their regenerative capacity over time and fall into decline. We know many examples from our own history such as the economic decline of the London Docklands and the pit communities in the north, or social decline in the major cities such as Manchester's Hulme district.

Sustainable urban regeneration requires a balanced mix of social, economic and environmental factors to enable each area to establish its own continuous regenerative capacity.

Urban regeneration projects tend to be large scale and multi-faceted; successful schemes incorporate mixed uses such as commerce, retail, industrial and of course residential. They also have a mixture of tenures including freehold and leasehold in both the public and private sectors. Projects often involve many different organisations such as local authorities, government bodies, private developers or house builders and social housing providers and more.

Key players

The key players in urban regeneration can be divided into the following groups:-

- Government Departments and Agencies: English Partnerships is the national regeneration agency and reports to ODPM. Regional and local agencies include the likes of SEEDA, EEDA, London Development Agency, Sheffield One etc.
- Local and regional authorities are often key landowners and major drivers of urban regeneration in their deprived areas. They also have land assembly powers.
- Housing associations, who are key to the provision of social and public sector housing and management.
- Private house builders are often essential to realising the economic value of many urban areas and providing initial capital investment.
- Private developers: many schemes require commercial, industrial, retail and other property essential to the initial and continuing regenerative capacity of an area.
- Private business and occupiers are essential to local economic welfare.
- Investors and funders are essential for raising the requisite capital.
- Infrastructure companies including transport, statutory supplies and services.

Types of regeneration

Urban regeneration can be classified into three main types:-

- Urban centre renewal – town centre regeneration schemes like Lewisham High Street.
- Estate regeneration – generally housing based schemes in deprived and declining areas.
- Urban renewal and new urban villages – schemes like the Greenwich Millennium Village.

Cost management

Key areas for growth

The ODPM has identified four key areas for growth to meet a massive predicted shortfall in housing provision in the South East. These include:-

- ❖ Thames Gateway
- ❖ Cambridge - Stanstead - London Link
- ❖ Milton Keynes (and East Midlands)
- ❖ Ashford.

The predominant focus in these areas will be to bring developers together with the central government agencies tasked to deliver government housing policy. This will be influenced by a switch of development responsibilities, particularly funding, away from RSLs direct to developers.

Procurement strategy

The projects associated with urban regeneration are large and activity is set to continue over the long term. Various forms of Public Private Partnership are essential in bringing together necessary resources and skills to undertake:

- ❖ land assembly and contamination issues;
- ❖ infrastructure development for transport and statutory services;
- ❖ social infrastructure development for essential schools, hospitals, affordable housing, etc;
- ❖ planning and urban design for sustainable communities and buildings;
- ❖ construction industry supply chain management, standardisation and innovation in techniques; and
- ❖ funding and capital development routes.

New skills will be required in integrating development management with project management in the context of regeneration.

Each regeneration project will have its own unique and challenging dimensions across the economic, social and environmental factors that make for sustainable placemaking for the benefit of people today and future generations.

VAT in the construction industry – zero rating

VAT and construction works

VAT in relation to buildings and construction is a complex area but below are some of the fundamental issues to consider when building new or carrying out works on an existing property:-

Building new property

All goods and services supplied by the industry for use in the construction of a building are standard rated (at 17.5% generally and 5% in specific situations) except in the following circumstances, when they would be zero rated:-

Where the new building is a dwelling fulfilling the following criteria:-

- ❖ Self contained.
- ❖ Able to be sold as a single dwelling.

Cost management

- ❖ Has been granted planning consent.
- ❖ Entitled to be used as a dwelling throughout the year.
- NB: Certain elements within a dwelling will always be standard rated (see later note). This zero rating only applies to goods and services supplied in the course of construction of a new building and to the first grant of a major interest, by the developer.

Relevant residential building fulfilling the following criteria:-
- ❖ Some facilities shared by residents, for example, children's homes, old people's homes, hospices, living quarters for school pupils or armed forces.
- ❖ An institution which is the sole or main residence for 90% of its residents.
- NB: Flats and sheltered housing schemes made up of individual flats will be zero rated as dwellings rather than relevant residential buildings.
- NB: Prisons, hospitals and hotels are specifically excluded from zero rating. This applies to goods and services and the first grant of a major interest.

Relevant charitable building is a building used solely by a charity for the <u>non business use</u> of the charity or as a village hall.
- NB: The supply must be made to the person who intends to use the building for such purposes before the supply is made then the person receiving it must give to the person making the supply a certificate that the intended use is for relevant residential or charitable purposes.

Works to an existing property

All goods and services supplied will be standard rated except in the following circumstances:-

Relevant Housing Association (HA) converting from non residential to residential use, ie the HA must:
- ❖ be a Registered Social Landlord within the meaning of Part 1 of the Housing Act 1996; and
- ❖ be a registered HA within the meaning of the Housing Associations Act 1985 (or Part II of the Housing (Northern Ireland) Order 1992).
- NB: This applies to goods and services only.

Approved alterations to existing protected buildings – basic principles as follows:-
- ❖ The work must be to a protected building as defined in VAT law and to the fabric of that building (for example, a listed building or scheduled monument).
- ❖ The work must require and be granted listed building consent.
- ❖ The works must not be of a repair or maintenance nature.
- ❖ The works must be apportioned if there is an element of both repair and maintenance and approved alteration.
- NB: Unlisted buildings in conservation areas do not qualify as protected buildings for the purposes of VAT relief. This applies to goods and services only.

Substantial reconstruction to existing protected buildings – basic principles as follows:-
- ❖ Definition of protected buildings as above.

Cost management

- ❖ The work must require and be granted listed building consent.
- ❖ At least 60% of the cost is attributable to 'approved alterations' or
- ❖ Only the wall(s) remain along with the other features of architectural or historic interest.
- NB: This applies to the first grant of a major interest to the person carrying out the substantial reconstruction.

Conversion of buildings from non-residential use to dwellings or relevant residential use:-

- ❖ From 1 March 1995 if you are the person converting a building the first grant of a major interest can be zero rated provided the building was neither designed nor adapted as a dwelling (or relevant residential purpose), or if it was designed as a dwelling and/or subsequently adapted, it has not been used as a dwelling for the whole of the previous ten years.

Residential conversions

From 12 May 2001 a lower rate of 5% applies with effect to the following supplies of building services and related goods:-

- ❖ A 'changed number of dwellings conversion', ie a conversion of a building (or part of a building) so that after conversion it has a different number of single-household dwellings (SHD) from the number before the conversion.
- ❖ A 'house in multiple occupation conversion', ie a conversion of a building (or part of a building) containing one or more single-household dwellings so that it contains only one or more multiple-occupancy dwellings.
- ❖ A 'special residential conversion', ie a conversion of premises containing dwellings (single-household or multiple occupancy) for use solely for a relevant residential purpose or converting a care home into a single-household dwelling.

Renovation and alteration of dwellings

The lower rate of 5% also applies to supplies of:

- ❖ building services; and
- ❖ related goods.

In the course of the alteration (including extension) or renovation of a single-household dwelling that has been empty for three years. 'Empty' means unlived in, so use for another purpose, such as storage, is acceptable. The dwelling can remain a single household.

Specific conditions apply to both of these lower rate situations.

The reduced rate of VAT

The reduced rate of 5% VAT applies from the 1 June 2002. The reduced rate is extended to the costs of:

- ❖ converting a non-residential property into a care home or multiple occupancy dwelling eg bedsits;
- ❖ converting a building used for 'relevant residential' purpose into a multiple occupancy dwelling;
- ❖ renovating or altering a care home or other qualifying building that has not been lived in for three years or more; and
- ❖ constructing, renovating or converting a building into a garage as part of the renovation of property that qualifies for the reduced rate.

Cost management

Charity annexes

Prior to 1 June 2002 all of an annexe had to be used or intended for use for a relevant charitable purpose in order to zero rate its construction. From 1 June 2002, Customs have issued a concession whereby minor non-qualifying use for example, use where the annex or part of an annexe is not used solely for a relevant charitable purpose, can be ignored with there being no requirement to apportion these between relevant charitable purpose and alternative usage.

Supplies to handicapped people

Certain goods and services supplied to a handicapped person, or to a charity for making these available to handicapped people for their domestic or personal use, may be zero-rated.

DIY projects

For claims made on or after 29 April 1996, Customs and Excise will refund any VAT chargeable on the supply, acquisition or importation of any goods used in connection with construction or conversion work where the following criteria are met. The work is not part of a business project; the work comprises the construction of a dwelling; relevant residential or charitable building; or is the conversion of a non-residential building into a dwelling.

All works, which fulfil the criteria set out above, can be zero-rated. However there are some items, within these categories, which will always be standard-rated:

- site investigations
- temporary site fencing
- concrete testing
- site security
- catering
- cleaning to site offices
- temporary lighting
- transport and haulage to and from site
- plant hire (without operator)
- professional services (architects, engineers, surveyors, solicitors etc)
- landscaping
- furniture (other than fitted kitchens) – material only
- some electrical or gas appliances – material only
- carpet and carpeting materials (including underlay and carpet tiles) – material only.

See the following notices issued by HM Customs and Excise for further expanded information and definitions:

- Notice 708 VAT Buildings and construction
- Notice 742 Land and property
- Notice 719 VAT refunds for 'do it yourself' builders and converters
- Notice 701/1 Charities
- Notice 701/7 VAT reliefs for people with disabilities
- Notice 701/19 Fuel and power
- Notice 708/5 Registered social landlords (housing associations etc)

Cost management

The primary legislation relating to this subject includes:-

- ❖ Group 5 of Schedule 8 to the VAT Act 1994 as amended by Statutory Instrument 1995 No 280 and Statutory Instrument 1997 No 50 – buildings under construction
- ❖ Group 6 of Schedule 8 to the VAT Act 1994 as amended by Statutory Instrument 1995 No 283 – approved alterations to protected buildings
- ❖ Group 12 of Schedule 8 of the VAT Act 1994 – supplies to the handicapped
- ❖ Section 35 of the VAT Act 1994 – DIY scheme
- ❖ The VAT (Input Tax) Order 1992 as amended by Statutory Instrument 1995 No 281
- ❖ Schedule 10 to the VAT Act 1994.

Capital allowances for taxation purposes

The following provides a brief summary of the capital allowances that are available to UK property owners, occupiers and investors. It is intended as an overview at the time of publication and the authors would direct readers who require more detailed or specific advice to contact a capital allowances expert.

The UK taxation system has no general provision for tax relief for capital expenditure or for the depreciation in value of capital assets with the passage of time. Although there are numerous accounting standards detailing how the depreciation of capital assets must be dealt with for accounts purposes, there is no allowable deduction for tax purposes. Accounting depreciation is therefore added back to the accounting profits when computing the tax charge.

Capital allowances provide tax relief for at least some of this accounting and tax mis-match. Qualifying capital expenditure incurred on certain buildings, fixtures and chattels will attract specific rates or amounts of tax relief, which are available to be offset against taxable profits.

This valuable form of tax relief is, in most cases, either under-claimed or not claimed at all due to a lack of understanding or application of the legislation and case law governing the availability of the relief.

General Scheme of Allowances

There are certain hurdles that must be cleared before capital allowances can be claimed. The entity incurring the expenditure must be within the charge to UK tax, therefore local authorities, government departments, charities, pension funds, etc will not be in a position to use the relief (although they may be able to pass the benefit on for consideration).

The expenditure must be capital in nature, ie there must be an enduring benefit to the trade of the entity incurring it, eg the initial acquisition of an asset, or the subsequent improvement of it. Expenditure on the maintenance or general upkeep of an asset or on the development of an asset held as trading stock is likely to be classed as revenue. Revenue expenditure will not qualify for capital allowances, but could still potentially be tax deductible in its entirety in the year it is incurred if it is written off through the profit and loss account. It should be noted that the revenue deduction is dependant on the accounting treatment of the expenditure. Revenue expenditure that is capitalised for accounts purposes, eg to improve the balance sheet, will not qualify for the 100% deduction in the year it is incurred.

Cost management

The capital expenditure must be incurred for the purpose of a qualifying activity, eg a trade, profession, vocation, etc. In some circumstances, the asset that is acquired or created must also be "in use" for the purpose of the qualifying activity.

Finally, the capital asset that is acquired or created must be qualifying for capital allowances purposes, and in this regard many tests have evolved through statute and case law.

Types and rates of allowances

Plant and Machinery Allowances (P&MAs)

These are given at a rate of 25% per annum on a reducing balance basis, ie 25% of the balance of the qualifying expenditure carried forward year on year. There is generally little help from the tax legislation on what constitutes qualifying P&M and therefore there has been much litigation in this area. Specialist advice should be taken, but typical examples will include IT equipment, heating and ventilation systems and specialist mechanical and electrical installations.

Qualifying P&M that has a useful economic life, when new, of exceeding 25 years is deemed to be a Long Life Asset (LLA) by the capital allowances legislation. The test applies from when the P&M is first used, therefore a second hand asset with a useful economic life of less than 25 years at the date of purchase may still be a LLA in the new owners business. LLAs attract allowances at a rate of 6% per annum on a reducing balance basis and are therefore less attractive than non-LLAs. Excluded from this is expenditure incurred on a dwelling house, hotel, office, retail shop or showroom.

Expenditure incurred on new (unused) qualifying P&M by a small or medium sized enterprise (SME) will attract a first year allowance of 40%. The balance will be relieved in subsequent years on the 25% reducing balance basis.

Specific legislation exists to give a 100% first year allowance for expenditure incurred on new P&M that satisfies defined energy saving or environmentally friendly criteria.

Hotel Building Allowances (HBAs), Industrial Buildings Allowances (IBAs) and Agricultural Buildings Allowances (ABAs)

These categories of buildings have been grouped together as they all attract allowances at the rate of 4% per annum on a straight-line basis, ie the qualifying expenditure is written off over a period of 25 years from when the building is first used.

The properties must satisfy certain criteria specified in the capital allowances legislation in order to be "qualifying". For HBAs, the property must be open for four months or more between April and October, have a minimum of 10 letting bedrooms and provide guest services of at least breakfast, dinner and room cleaning. For IBAs and HBAs the test relates to the use to which the building is put, therefore there has to be qualifying industrial or agricultural use for the relief to be forthcoming.

Research and Development Allowances (R&DAs)

Expenditure incurred on facilities used for qualifying R&D attracts allowances at 100% in the first year. Again, the accounting treatment of the expenditure is key and this is supplemented by published guidelines providing more clarity. The underlying test is innovation.

Capital allowances issues for property transactions

Except where it has already been identified that qualifying expenditure must be on "new" P&M, capital allowances are available on the purchase of second hand assets. There are many tests and restrictions that apply to

Cost management

property transactions to ensure that a subsequent owner is restricted in its claim to the original cost of the asset. Where no previous capital allowances claim has been made, there may be an opportunity to "step up" the new owners claim to reflect the price that it has paid for the asset, where this is greater than the original cost.

The vendor and the purchaser must apportion the purchase price, on a just and reasonable basis, between the components of the sale, i.e. the interest in the land, the building and the qualifying P&M. These figures are then compared to the original cost to establish what restriction may apply to the subsequent claim and conversely what disposal proceeds the vendor must account for on sale. In some situations, it is possible for the parties to the transaction to agree and formally elect for a specified value to be attributable to fixtures qualifying for P&MAs. The effect of the election is that all, part or none of the allowances can be passed between the transacting parties. Tax aware parties may be able to negotiate an adjustment to the purchase price to reflect the agreed capital allowances position.

It is important to establish the capital allowances history of a property during the pre-acquisition due diligence. The information that will be relevant to the claim is generally more readily made available at this time, than post-transaction and the opportunity to obtain any additional benefit through joint elections may not arise once the deal is concluded.

The following table provides a guide to the expected levels of qualifying P&M that may be available on an unrestricted basis for differing property and transaction types. It should be noted that depending on the type of property, ABAs, HBAs or IBAs might also be available in addition to the expected P&MAs identified.

Property type	Acquisition / disposal Qualifying %	New build Qualifying %
Office	14% - 25%	15% - 45%
Office refurb/fit-out	-	40% - 100%
Retail	2% - 20%	5% - 35%
Retail refurb/fit-out	-	40% - 90%
Industrial	2% - 15%	5% - 20%
Hotel	10% - 40%	15% - 50%
Hotel refurb/fit-out	-	40% - 80%

Summary

The above is intended as an overview of the current availability of capital allowances. Specialist advice should be sought to maximise the availability and quantum of capital allowances.

Further sources of information

Capital Allowances Act 2001

Income and Corporation Taxes Act 1988

Finance Act 2004

www.eca.gov.org

Land remediation allowances

Various tax incentives have been introduced following the recommendations of the Urban Task Force set up by the government in 1998 to investigate causes of urban decline. Land Remediation Tax Relief is one such generous incentive that provides up to 150% of tax relief for expenditure incurred in remediating contaminated land or, where a company is loss making, a tax credit (payment from the Exchequer) of 24% of the qualifying cost.

Availability of tax relief

The qualifying expenditure must be incurred after 11 May 2001 on qualifying remediation works to land situated in the UK that was acquired for the purpose of a trade in a contaminated state. Land includes interests over land and buildings on the land.

The relief is available for companies but not for individuals or partnerships. A company in limited partnership can claim the relief in respect of its share in the partnership's expenditure on remediation. The relief is not available where the land is contaminated due to the actions of the acquiring company or someone connected with that company, or where the company has failed to prevent contamination.

Land is in a contaminated state if there are substances in, on or under the land that are likely to cause harm to people, property or ecological systems or is likely to pollute controlled waters. Nuclear sites are specifically excluded from the definition of contaminated land.

Qualifying expenditure

Expenditure must be incurred on the prevention, remediation or mitigation of the effects of the pollutant or on the restoration of the land to its former state. The expenditure must be directly linked to the remediation and as such, general site clearance will not qualify. Where in-house employees undertake the work, all of these costs are qualifying provided at least 80% of the time is spent on remediation tasks. Where employee time is less than 80%, some pro-rata adjustment will be necessary.

Preparatory works such as site investigations and incidental professional fees can be included in the claim for the relief. Where remediation work is sub-contracted, then this cost will form the basis of the claim.

Claiming the relief

This relief is available to property owners, investors and developers. Any claim for the relief must be made within two years of the end of the accounting period in which the qualifying expenditure was incurred.

Interaction with landfill tax and aggregates levy

There are tax-planning opportunities to structure projects such that the waste from qualifying remediation works is exempt from landfill tax or the aggregates levy.

Summary

The above is intended as an overview of the principles and the current availability of land remediation tax relief. Specialist advice should be sought on the availability and to maximise the quantum of the tax relief.

Further sources of information

Income and Corporation Taxes Act 1988
Finance Act 2001
www.inlandrevenue.gov.uk

Cost management

Tender prices and building cost indices

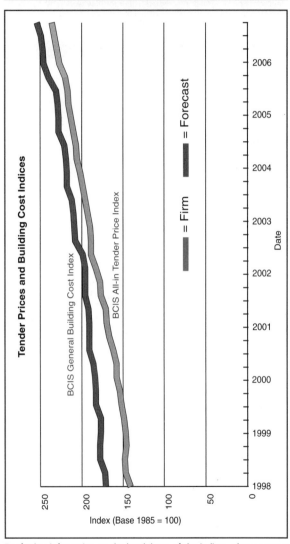

For further information on the breakdown of the indices, please contact the Building Cost Information Service.

Cost management

The BCIS tender price indices and general building cost indices monitor the movement of tender prices and building costs. They can be used to forecast cost movement for the years to come by using the forecast indices based upon cost trends. The Tender Price Index measures the trend of contractors pricing levels in accepted tender for new work (cost to client), whereas the General Building Cost Index measures changes in costs of labour, materials and plant (cost to contractor).

Rating revaluation 2005

Introduction

Since the introduction of the Local Government Finance Act 1988, rating assessments on premises are subject to five yearly revaluations. The Valuation Office is now preparing the new 2005 Rating List, which will come into affect on 1 April 2005.

Basis of valuation

The 2005 Rateable Values will be based on rental values on 1 April 2003, which is known as the Antecedent Valuation Date (AVD). The rating hypothesis as set out in the Local Government and Finance Act 1988 will continue to form the basis of valuation.

Right first time

The Valuation Office Agency (VOA) intends to approach the 2005 revaluation under a new modernised agenda to be 'right first time'. The Government felt that past revaluations encouraged too many appeals and wanted to create a more efficient system to more accurately forecast budgets and to ensure that the Uniform Business Rate (UBR) is set at an appropriate level.

'Notice Requesting Supply of Information...'

In order to be 'right first time' the VOA will require the return of Information Notices setting out rental information on any given property. The proportion of forms returned is low and the new Local Government Bill will include a clause that entitles the VOA to issue a fine of £100 if the form is not completed and returned within 56 days of its issue, to be increased by £20 per day thereafter.

Uniform Business Rate

Currently, the UBR increases by the rate of inflation for each rate year. However, the Government proposes powers to amend the UBR for each rate year as a fail-safe measure to claw back additional rates once appeal activity begins to slow.

Transitional relief

The indications are that the Government intends to retain transitional arrangements to cushion the impact of increases and the benefits of reductions on the basis that businesses need to be able to budget. However, the arrangements will not be announced until Autumn 2004.

The format

Following a campaign by RICS, the transitional arrangements will be spread over the five year period of the List to avoid sharp increases over the first few years. Downward phasing of the transitional relief will continue to apply to recoup the losses incurred by the government.

Small business relief

The Bill further proposes a relief scheme for properties of a rateable value of below £8000. The relief will vary between 0-50% on a sliding scale dependent upon the rateable value, although this will only be available to businesses that occupy only one such property.

Timetable - 2005 Revaluation

1 April 2003	-	The Antecedent Valuation Date for 2005 Rating assessments
Summer 2003	-	Local Government Bill likely to be enacted
Summer 2004	-	Preliminary announcement from the Government on Uniform Business Rate for 2005 and possible scheme of transitional arrangements
30 September 2004	-	Draft Revaluation assessments published and summary valuations issued to some rate payers
October 2004	-	Confirmation of UBR and transitional arrangements for 2005
1 April 2005	-	New 2005 List Rateable Values come into effect.

What can be done now?

Rental transactions such as lettings, rent review and lease renewal settlements close to the Antecedent Valuation Date of 1 April 2003 are likely to influence the level of the 2005 Rateable Value.

However, the likely level of rate liability for the 2005 List cannot be assessed until such time as both the Uniform Business Rate and transitional arrangements are announced, which is likely to be in Summer 2004.

When the Draft Rating List is published, by no later than 20 September 2004, consideration can then be given to the level of assessment and as to whether the assessment should be appealed when the list comes into effect on 1 April 2005.

Appeal process

There are likely to be some changes to the appeal procedures, although generally provided an appeal is made within a specified period, the new List will ensure that the backdating benefit of any reduction is to the effective date of the list.

Cost management

Predicting costs

Be mindful:

- ❖ of the level of information;
- ❖ of the perceived requirements for budget, time and quality;
- ❖ that the brief will inevitably change;
- ❖ that not all of your assumptions will be right; and
- ❖ that the initial assessment will stick in the client's mind.

Always:

- ❖ undertake a detailed assessment;
- ❖ set down clearly all assumptions and exclusions; and
- ❖ make provision for risk – in the market, in design development and in the client changing his/her mind.

Never:

- ❖ rely solely on blanket unit rates; or
- ❖ expect the client to remember anything other than your first assessment.

Cost management

The procurement process

Be mindful:

- that the process of project delivery is dynamic; and
- of the inevitability of the design (in part at least) developing apace with construction and that if this design development is not managed, cost and time over-runs will result.

Always:

- select an appropriate form of contract and consider not just the value but the nature of the work;
- understand the budget and know where the uncertainties are;
- maintain projections of cost on an 'open book' basis;
- make the projections 'real time' and as soon as issues are suspected, make provision; and
- assess and reassess risk.

Never:

- avoid routine financial appraisals;
- proceed on the basis that things will turn out all right if left to their own devices – they never will; or
- proceed without adequate contingency.

Value for money

Be mindful that:

- value for money is not just the lowest price;
- it must balance time, cost and quality; and
- simply applying competitive tender procedures does not itself demonstrate value for money.

Life cycle costing

The life cycle cost of an asset may be defined as the total cost of that asset over its operating life including initial capital/acquisition cost, occupation costs, operating costs, and the cost or benefit of the eventual disposal of the asset at the end of its life.

Use

Life cycle costing is essential to effective decision-making in four main ways:-

- It identifies the total cost commitment undertaken in the acquisition of any asset.
- It facilitates an effective choice between alternative methods of achieving a stated objective, recognising different patterns of capital and running costs.
- It is a management tool that details the current operating costs of assets.
- It identifies those areas in which operating costs might be reduced.

The use of life cycle costing techniques has seen considerable growth in recent years, driven predominantly by shifts in public sector procurement policy towards 'best value' rather than lowest cost. This policy change is demonstrated in publications such as 'Construction Procurement

Cost management

Guidance, No 7 Whole Life Costs' (Office of Government Commerce) which states that "all procurement must be made solely on the basis of value for money in terms of the optimum combination of whole life costs and quality to meet the user's requirements". The increasing reliance upon PFI / PPP for the procurement of major capital projects has also contributed to the need to consider the whole life cost of an asset rather than simply the 'up front' capital outlay.

The benefits of life cycle costing should not be construed as being effective for the public sector alone – long term investment returns on property can be influenced significantly by future maintenance/operational obligations which were not considered at the outset of a project. Energy efficiency of buildings in particular, is considered to be a prime candidate for life cycle cost evaluation. In an era dominated by European directives, the Climate Change Levy etc, the higher capital cost of energy efficient installations can often be more than offset by longer term savings. Adding the benefit of potential tax advantages, the value of life cycle costing becomes apparent.

As a result, life cycle cost planning should, wherever possible, form an integral part of the design development process, being implemented from project outset to achieve maximum benefit.

Implementation

Implementation of appropriate life cycle costing techniques must reflect not only the nature of the client's needs but also their property specific objectives. The objectives of an owner occupier client, retaining a long term interest, will be significantly different from an investor seeking to transfer maintenance risk to a prospective tenant.

The following major elements should therefore be considered:-

- ❖ The overall time period.
- ❖ All costs and revenues attributable to the project, including initial investment, recurring costs and revenues, proceeds from ultimate sale or other disposal and tax benefits.
- ❖ Only those costs and revenues directly attributable to the project.
- ❖ The effects of time, including allowance for the impact of inflation.
- ❖ The fact that pounds spent or received in the future are worth less than pounds spent or received today.

Stages of life cycle costing

Life cycle costing techniques can be split into the following stages :

Life cycle cost planning (LCCP)

Objectives:-

- ❖ To identify the total costs of the acquisition of a building or building element.
- ❖ To facilitate the effective choice between various methods of achieving a given objective.

LCCP can be broken into seven basic steps:-

Step 1 - Establish the objective.

Step 2 - Choose alternative options for achieving the objective (consider all realistic possibilities).

Step 3 - Formulate assumptions (eg discount rate, construction duration, asset replacement cycle, building life, basis of residual value).

Step 4 - Calculate all capital and recurrent costs over a defined period.

Cost management

The sequence linking LCCA, LCCP and LCCM

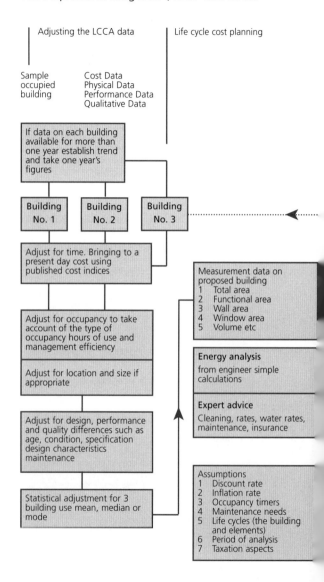

Cost management

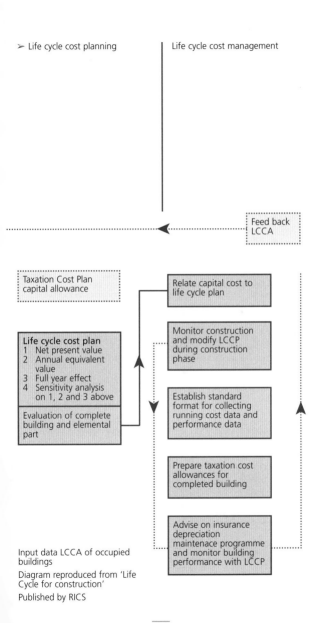

Cost management

Step 5 - Compare costs and rank the alternatives.

Step 6 - Carry out a sensitivity analysis (ie when the results of step 5 are not demonstrably in favour of one choice).

Step 7 - Investigate capital cost constraints.

Applying the foregoing approach for each option under consideration will generate a life cycle cost plan, reflecting all costs and revenues over the selected time period. Recognising that monies spent in the future will be worth less than the same money spent today, the cash flows are discounted to a present day value. The total of all present day values over the life of the LCCP represent the net present value (NPV). Compared against alternative options, the NPV facilitates identification of the option which offers best value for money.

Life cycle cost management (LCCM)

Objectives:-

- ❖ To identify areas where running costs might be reduced.
- ❖ To establish where performance differs from the LCCP projection.
- ❖ To make recommendations on more efficient utilisation of the building.
- ❖ To provide information on asset lives and reliability factors for accounting purposes.
- ❖ To assist in the establishment of a maintenance policy for the building.
- ❖ To give taxation advice on building related items.

LCCM therefore applies throughout the operational phase of a building's life, focused upon maintaining and/or improving the asset's efficiency. LCCM should be read in conjunction with maintenance management.

Life cycle cost analysis (LCCA)

Objectives:-

- ❖ To provide a management tool which identifies actual operating costs incurred; and
- ❖ To relate cost and performance data to advisers about the running costs of occupied buildings.

LCCA is an ongoing operation, contributing to both LCCP and LCCM – in practice, one cannot take place without the other.

While each component can be viewed as a separate activity in its own right, there is a logical sequence that links them together. The sequence is demonstrated in figure 1, where a life cycle cost plan for a proposed project is based upon three similar buildings for which life cycle cost analyses are available.

Main components

Consider the following when implementing life cycle cost techniques:

1 Capital costs
- Land (including cost of acquisition)
- Demolitions
- Construction costs
- Fit-out expenditure
- Decanting charges (incl. rental of temporary accommodation)
- Relocation expenses
- Professional fees
- Statutory fees
- VAT (and other taxes)

Cost management

2 Financing charges	• Cost of finance for land/site acquisition • Cost of finance during occupational period
3 Operation costs	• Heating, electricity and other energy • Cleaning and caretaking • Rates • Insurance • Security and public health • Staff (building related) • Rent • Management charges • Land charges
4 Maintenance costs	• Life cycle asset replacement (specific) • Annual maintenance (general) • Annual grounds maintenance (general)
5 Residual values	• Disposal of existing buildings, land etc • Disposal of temporary decant accommodation • Retained value at end of evaluation period
6 Revenue	• Other income derived from the accommodation • Benefit of capital allowances or other incentive.

Other considerations

The discount rate and inflation

When compiling life cycle cost plans, it is important to distinguish between the interest rate and the discount rate. Where project cash flows include an allowance for inflation, the discount rate should make an equivalent allowance. Alternatively, where estimates are in current prices, they should be discounted at the real discount rate.

While there is no definitive approach, it is reasonable to state that if all cost estimates are expected to inflate at the same rate then it is not unreasonable to perform all calculations in current prices, applying a real discount rate. Alternatively, where cost estimates are likely to inflate at different rates (eg energy costs inflating significantly faster than construction prices) than it may be beneficial to account for the differing rates of inflation.

In the case of financing through borrowing, the discount rate will equal the long term cost of borrowing money in the market place (net of inflation). While the discount rate will vary with the source of funding, it is worth noting that HM Treasury apply a discount rate of 3.5% in public sector option appraisal.

Project life

An estimate of the probable life of the project should be made. When ranking projects with identical lives, the option with the lowest present value should be chosen. However, where the options have different lives, the cost of each alternative should be expressed as an annual equivalent and the option with the lowest annual equivalent chosen.

Reinstatement valuations for insurance purposes

Generally it is wise to insure a property for the value of complete reinstatement following total destruction (including partial destruction that necessitates demolition and rebuild). The following points should be considered:-

- ❖ Demolition cost to be valued.
- ❖ Rebuilding costs to be valued (initially at 'Day One' rates). Care must be taken here to include for any special features that are to be included to enable replacement of the property on a like for like basis. Examples include façade treatments such as stonework embellishments or internal features such as specialist decorations.
- ❖ Extent of the valuation to be discussed and agreed with the client: whether tenants fixtures and fittings are to be valued; whether a basic landlord fit out is to be included; or a shell only valuation with all other items to be included under separate policies. This may be done on agreement with the client or by reference to leases which state the obligations of the Landlord and Tenant.
- ❖ Professional fees to be included at an appropriate level, depending on the complexity and location of the property.
- ❖ Geographical location factors to be accounted for either in re-building rates or separately by reference to recognised indices.
- ❖ Local authority planning and building regulation fees must be included as they are an unavoidable expense.

If the valuation is being projected to cover the period of the policy (as opposed to Day One valuation) then account should be made for the:

- ❖ period of the policy;
- ❖ design period;
- ❖ planning period;
- ❖ construction period; and
- ❖ void (letting) period (if required by the owner).

This may result in projecting costs for anything up to and beyond a three year period. In addition, as with any construction scheme, a contingency may be included.

In summary, a Day One valuation may increase by upwards of 30% on the basic rebuild costs to take account of the issues listed above. Property owners should always seek a realistic valuation of their property to avoid being over or under-insured. At best the premiums may be based on an exaggerated value and at worst, following destruction, there will be insufficient funds to reconstruct.

Technical resources

- Useful information sources — 254
- Useful website addresses — 255
- Contributors to the Watts Pocket Handbook — 260
- Watts and Partners' publications — 264

Technical resources

Useful information sources

Association for Project Management
Tel: 08454 581944 Web: www.apm.org.uk

Association of Building Engineers
Tel: 01604 404121 Web: www.abe.org.uk

Association of Consultant Approved Inspectors
Tel: 01435 862487 Web: www.acai.org.uk

Building Cost Information Service Ltd
Tel: 020 7695 1500 Web: www.bcis.co.uk

Building Maintenance Information
Tel: 020 7695 1500 Web: www.bcis.co.uk

Building Research Establishment
Tel: 01923 664000 Web: www.bre.co.uk

British Cement Association
Tel: 01276 608701 Web: www.cementindustry.co.uk

British Institute of Facilities Management
Tel: 01799 508606 Web: www.bifm.org.uk

British Property Federation
Tel: 020 7828 0111 Web: www.bpf.org.uk

British Standards Institution
Tel: 020 8996 9000 Web: www.bsi-global.com

Building Services Research and Information Association
Tel: 01344 465600 Web: www.bsria.co.uk

Centre for Accessible Environments
Tel: 020 7840 0125 Web: www.cae.org.uk

Centre for Window and Cladding Technology
Tel: 01225 386541 Web: www.cwct.co.uk

Chartered Institute of Arbitrators
Tel: 020 7421 7444 Web: www.arbitrators.org

Chartered Institute of Building
Tel: 01344 630700 Web: www.ciob.org.uk

Commission for Architecture and the Built Environment
Tel: 020 7960 2400 Web: www.cabe.org.uk

Concrete Repair Association
Tel: 01252 739145 Web: www.concreterepair.org.uk

Concrete Society
Tel: 01276 607140 Web: www.concrete.org.uk

Construction Health and Safety Group
Tel: 01932 561871 Web: www.chsg.co.uk

Construction Industry Council
Tel: 020 7399 7400 Web: www.cic.org.uk

Construction Industry Research and Information Association
Tel: 020 7549 3300 Web: www.ciria.org.uk

Design Council
Tel: 020 7420 5200 Web: www.designcouncil.org.uk

Disability Rights Commission
Tel: 08457 622633 Web: www.drc-gb.org

Energy Watch
Tel: 08459 06 07 08 Web: www.energywatch.org.uk

English Heritage
Tel: 08703 331181 Web: www.english-heritage.org.uk

Environment Agency
Tel: 08708 506506 Web: www.environment-agency.gov.uk

Fire Protection Association
Tel: 01608 812500 Web: www.thefpa.co.uk

Glass & Glazing Federation
Tel: 0870 0424255 Web: www.ggf.org.uk

Health and Safety Executive
Tel: 08701 545500 Web: www.hse.gov.uk

HM Land Registry
Tel: 020 7917 8888 Web: www.landreg.gov.uk

House Builders Federation
Tel: 020 7960 1600 Web: www.hbf.co.uk

Institution of Civil Engineering Surveyors
Tel: 0161 972 3100 Web: www.ices.org.uk

Institution of Structural Engineers
Tel: 020 7235 4535 Web: www.istructe.org.uk

Joint Contracts Tribunal
Tel: 020 7637 8650 Web: www.jctltd.co.uk

Lead Development Association
Tel: 020 7499 8422 Web: www.ldaint.org

Mastic Asphalt Council
Tel: 01424 814400 Web: www.masticasphaltcouncil.co.uk

National Trust
Tel: 0870 609 5380 Web: www.nationaltrust.org.uk

Office of the Deputy Prime Minister
Tel: 020 7944 4400 Web: www.odpm.gov.uk

RIBA Bookshop
Tel: 020 7256 7222 Web: www.ribabookshop.com

Royal Forestry Society
Tel: 01442 822028 Web: www.rfs.org.uk

Royal Institute of British Architects
Tel: 020 7580 5533 Web: www.riba.org

The Royal Institution of Chartered Surveyors (RICS)
Tel: 0870 333 1600 Web: www.rics.org

The Royal Institution of Chartered Surveyors in Scotland
Tel: 0131 225 7078 Web: www.rics-scotland.org.uk

RICS Books
Tel: 0870 333 1600: www.ricsbooks.org

Society of Construction Law
Tel: 01235 770606 Web: www.scl.org.uk

Stone Federation Great Britain
Tel: 01303 856123 Web: www.stone-federationgb.org.uk

The Society of Expert Witnesses
Tel: 0845 702 3014 Web; www.sew.org.uk

The Stationery Office Bookshop
Tel: 0870 600 5522 Web: www.clicktso.com

The Survey Association
Tel: 01784 223760 Web: www.tsa-uk.org.uk

Timber Research and Development Association
Tel: 01494 569600 Web: www.trada.co.uk

Timber Trade Federation
Tel: 020 7839 1891 Web: www.ttf.co.uk

Useful website addresses

Barbour Expert-online to the built environment
www.barbourexpert.co.uk
Well-known provider of information services to construction industry professionals and those responsible for health and safety at work.

Technical resources

Boundary Problems
www.boundary-problems.co.uk
This web site has been designed to increase understanding of property boundary and related issues. It aims to help minimise the difficulties that have been experienced by so many people who have suffered a neighbour dispute relating to their property boundary.

BRE Certification Ltd
www.brecertification.co.uk
A leading research-based consultancy, certification and testing business – covering the built environment and associated industries.

BREEAM
www.products.bre.co.uk/breeam
The BRE Environmental Assessment Method for assessing the environmental performance of both new and existing buildings.

The British Council for Offices
www.bco-officefocus.com
The British Council for Offices' mission is to research, develop and communicate best practice in all aspects of the office sector.
It delivers this by providing a forum for the discussion and debate of relevant issues.

British Geological Survey
www.bgs.ac.uk
Site covering details of products and services including building stones services for construction and conservation and an address-linked geological inventory.

Building
www.building.co.uk
UK industry magazine covering the whole spectrum of design and construction activity and all project disciplines.

Building Conservation
www.buildingconservation.com
Online information centre for the conservation and restoration of historic buildings, churches and garden landscapes.

Businessparks.net
www.businessparks.net
Subscription service providing up to date and in depth data on the UK business park market. The database is continuously updated by a specialist research team.

Compliance+
www.ciria.org/compliancplus/
Site giving guidance on environmental good practice in construction.

Contamlinks
www.contamlinks.co.uk
Portal containing over 500 links to a range of contaminated land and associated resources.

Construction Best Practice Program
www.constructing_excellence.org.uk
The Construction Best Practice Programme provides support to individuals, companies, organisations and supply chains in the construction industry seeking to improve the way they do business: clients, contractors, consultants, specialists, large or small, public or private. The programme offers a range of services, which raise awareness, gain commitment, support action and facilitate sharing.

Construction Europe
www.construction-europe.com
Site for the pan-European magazine for the construction industry including articles from the latest issues, links, a bookshop and events.

Technical resources

Construction Line
www.constructionline.co.uk
Constructionline is the UK's largest register of pre-qualified construction services. It streamlines procedures by supplying the construction industry and its clients with a single national pre-qualification scheme.

Construction Plus
www.constructionplus.co.uk
Portal site with links to news, companies and journals across all construction disciplines.

Construction Research and Innovation Strategy Panel (CRISP)
www.ncrisp.org.uk
CRISP brings together government, clients, industry and the research community to consider research priorities for construction. It has close links with other rethinking initiatives, including the Movement for Innovation, the Housing Forum and the Construction Best Practice Programme.

Construct Sustainability
www.constructsustainably.co.uk
In the increasingly competitive field of construction, sustainability is becoming a differentiator, capable of providing a vital competitive edge. This website is specifically intended to help the smaller constructors compete successfully in this new area alongside the bigger contracting organisations. It provides practical advice and information in answer to frequently asked questions on sustainability.

Corrosion Prevention Association
www.corrosionprevention.org.uk
The CPA's objective is to identify, quantify and communicate the principles and benefits of cathodic protection to those who have an interest in or influence the design, construction, use, maintenance, preservation and investment in reinforced concrete buildings and structures.

Countyweb
www.countyweb.co.uk
County by county guide to total information including a free property-listing service.

Court Service
www.courtservice.gov.uk
The Court Service is an executive agency of the Lord Chancellor's Department, whose purpose is the delivery of justice. The website covers policy, legislation and the Magistrates' courts.

DDA Tax Guidance
www.inlandrevenue.gov.uk/specialist/disability-act-guidance.htm
New tax guidance on tax allowances available to businesses to help them make their premises more accessible.

DfES Public Private Partnerships
www.dfes.gov.uk/ppppfi
Includes details of PFI projects in the schools sector and related publications.

DRA UK Construction Industry Links
www.dragroup.com
Database containing UK construction industry related links, categorised in 158 service categories allowing each category to be searched nationally, in 11 regional areas or in 124 postal areas.

Environmental Organisation Web Directory
www.webdirectory.com
Described as the 'Earth's biggest environment search engine'. Lists sites under 30 major headings.

Technical resources

The European Forecasting Group for the Construction Industry
www.euroconstruct.org
Forecasting trends for the construction market in Europe. It includes 19 European economic and technical research institutes. The site includes details of events and publications.

HM Treasury - PFI
http://www.hm-treasury.gov.uk
Government website dealing with policy issues relating to PFI. Follow Public Private Partnership link from this homepage.

The Housing Forum
www.constructingexcellence.org.uk
The purpose of the Housing Forum is to bring together parties involved in the house building supply chain who are committed and ready to become part of a movement for change and innovation in construction and renovation.

Information for Industry
www.connectingbusiness.com
An environmental business news briefing, environmental business magazine, environment business directory and a compliance manual are included on this site.

InsideEnergy
www.utilityweek.com/insideenergyv1/default.asp
An e-journal for energy in building and industry. Includes an update on the latest industry news, features, case histories, a comprehensive product directory, industry events and a forum to express and exchange news on industry topics.

Integrated Facilities Management
www.i-fm.net
An online resource for facilities management professionals.

Landlord Zone
www.landlordzone.co.uk
An online portal for landlords providing free access to information and resources of value to residential and commercial landlords, tenants, letting agents and other property professionals.

National Green Specification
www.greenspec.co.uk
NGS is partnered with the BRE to produce an internet resource for all involved with sustainable construction.

National Radiological Protection
www.nrpb.org
Website of the National Radiological Board. The site gives advice and information on Radon protection in the UK.

NetRegs
www.environmental-agency.gov.uk/netregs/
Website designed to help business to understand the complex regulations affecting their environmental obligations.

Public Private Finance
http://www.publicprivatefinance.com/pfi/
Private sector PFI news and market information database (subscription only).

The PPP Forum
http://www.pppforum.com/
Private sector website established to promote PFI and containing technical and market information.

Technical resources

Pyramus & Thisbe
www.partywalls.org.uk
An organisation for professionals specialising in party wall matters with news, events, FAQs, information about the organisation and a database of members. The website also provides information on the Party Wall etc Act 1996 and its practical application.

The Quango Website
www.cabinet-office.gov.uk/agencies-publicbodies
www.publicappointments.gov.uk
Site that lists quangos by government department or name of organisation and lists the names of people involved.

Radon Centres Ltd
www.radon.co.uk
Organisation that can help with domestic and commercial properties for the sale of Radon detectors, Radon testing surveys and detector placement and installation of Radon reduction systems.

Service Charges Guide
www.servicechargeguide.propertymall.com
This guide sets out overall principles for good practice. It is designed to cover commercial property.

SIMAP
www.simap.eu.int/
EU public procurement information website home page.
http://simap.eu.int/EN/pub/src/welcome.htm
SIMAP web pages with links to EU official forms, Common Procurement Vocabulary (CPV) and Territorial Nomenclature (NUTS).

Society of Property Researchers
www.sprweb.co.uk
Site for this professional grouping of researchers. Includes details of events, special interest groups and members.

Tenders Electronic Daily (TED)
http://ted.publications.eu.int/official/
The online Supplement to the Official Journal with a searchable database of all public tenders.

UK Online Government
www.ukonline.gov.uk
A portal site with links to all government online information and services.

UK Taxation Directory
www.uktax.demon.co.uk
An independently compiled catalogue of websites and material of potential interest to tax professionals and others seeking online information on UK tax matters.

Whole Life Cost Forum
www.wlcf.org.uk
The whole life cost forum is a collaborative initiative by all sectors of the construction industry. The forum has been set up to form the UK database of whole life cost and performance data.

Technical resources

Contributors to the Watts Pocket Handbook

External contributors

The scope of the Watts Pocket Handbook would not be possible without the cooperation of external professionals who have dedicated their expertise and valued time to produce specialist information for inclusion in the publication. For this help and support, Watts and Partners would like to express their sincere appreciation to the following:

Legal and lease
Adjudication under the Scheme for Construction Contracts - how to get started
Dispute resolution
Expert witness
The Provisions of Part II of the Housing Grants,
Construction and Regeneration Act 1996

Suzanne Reeves
Wedlake Bell
52 Bedford Row
London WC1R 4LR
Tel: 020 7395 3168
Fax: 020 7395 3004
Email: sreeves@wedlakebell.com

Building services design
Air conditioning systems
Data installations
Lift terminology
Lighting design
Plant and equipment
Conservation
Energy conservation
Environment and specification

Keith Crosby
Watkins Payne Partnership
51 Staines Road West
Sunbury-on-Thames
Middlesex TW16 7AH
Tel: 01932 781641
Fax: 01932765590
Email: keithcrosby@wppgroup.co.uk

Conservation
BREEAM

Mark Worthington
Tooley and Foster Partnership
19-20 Grosvenor Street
London W1K 4QH
Tel: 020 7290 9580
Fax: 020 7499 7566
Email: mworthington@tooleyfoster.com

Contracts and procurement
Procurement and standard form contracts

Joe Bellhouse
Wedlake Bell
52 Bedford Row
London WC1R 4LR
Tel: 020 7395 3073
Fax: 020 7406 1602
Email: jbellhouse@wedlakebell.com

Cost management
Capital allowances for taxation purposes
Land remediation allowances

Chris Doyle
Yewell Consultancy
1 Hall Road
Wallington
Surrey SM6 0RT
Tel: 020 8395 0397
Fax: 020 8395 0397
Email: chrisdoyle@yewell-consultancy.co.uk

Technical resources

Cost management
Rating revaluation 2005
Legal and lease
Section 18(1) of the Landlord and Tenant Act 1927

David Shorthall
Alexander Reece Thomson
11 Welbeck Street
London W1G 9XZ
Tel: 020 7486 1681
Fax: 020 7486 4200
Email: davidshorthall@arthom.co.uk

Cost management
Tender prices and building cost indices

Joe Martin
Building Cost Information Service
Royal Institution of Chartered Surveyors
3 Cadogan Gate
London SW1X 0AS
Tel: 020 7695 1500
Fax: 020 7695 1501
Email: bcis@bcis.co.uk

Cost management
VAT in the construction industry - zero rating

Adrian Houstoun
Kingston Smith
Devonshire House
60 Goswell Road
London EC1M 7AD
Tel: 020 7566 3802
Fax: 020 7566 4010
Email: ajh@kingstonsmith.co.uk

Design
Basic design data
Building types

Robert Clements
Tooley and Foster Partnership
19-20 Grosvenor Street
London W1K 4QH
Tel: 020 7290 9580
Fax: 020 7499 7566
Email: rclements@tooleyfoster.com

Site analysis
Measured surveys

Simon Barnes
Plowman Craven and Associates
141 Lower Luton Road
Harpenden
Hertfordshire
AL5 5EQ
Tel: 01582 765566
Fax: 01582 763180
Email: sbarnes@plowmancraven.co.uk

Town and country planning in England
Appeals and called-in applications
Development applications and fees
Development plans and monitoring
Environmental impact assessments
General Development Order and Use Classes Order
Listed buildings and conservation areas
Planning policy guidance notes and circulars

Malcolm Judd
Malcolm Judd and Partners
70 High Street
Chislehurst
Kent BR7 5AQ
Tel: 020 8289 1800
Fax: 020 8289 1200
Email: malcolm.judd@mjp.uk.net

Technical resources

Watts and Partners' Contributors

Access agreements	Aidan Cosgrave
Acts of parliament and regulations	Angela Dawson
Airtightness	Trevor Rushton
Asbestos	Paul Winstone, Trevor Rushton
'Best value' in local authorities	Stuart Russell
The Building (Amendments) Regulations 2004	Trevor Rushton
Chemical and physical testing requirements	Trevor Rushton
Chlorides	Trevor Rushton
Commercial/industrial surveys	Trevor Rushton
Common defects in commercial and residential properties	Trevor Rushton
Composite panels	Trevor Rushton
Computer-aided design	Stephen Kent
Condition surveys	Steve Brewer
The Construction (Design and Management) Regulations 1994	Paul Winstone
Construction (Health, Safety and Welfare) Regulations 1996	Paul Winstone
Construction management	Mike Ridley
Construction noise and vibration	Aidan Cosgrave
Contaminated land	Tim Coffin
The contract administrator's role	Andrew Gear
Contributors to the Watts Pocket Handbook	Samantha Rumens
Conversion formulae	Angela Dawson
Coordinating project information	Daniel Webb
Corrosion of metals	Trevor Rushton
Cost management	Christopher Knott
Curtain walling systems	Trevor Rushton
Daylight and sunlight amenity	Aidan Cosgrave
Defects in concrete	Trevor Rushton
Deleterious materials	Trevor Rushton
Design loadings for buildings	Trevor Rushton
The design process	Graeme Lees
Development monitoring	Campbell MacDougall, Rebecca Jermy
Different specifications for different contracts	Tim French
Dilapidations	Alex Charlesworth
Disability Discrimination Act 1995	Jon Hubbard, Mark Ratcliff
Discovery of building defects - statutory time limits	Paul Lovelock
Due diligence and the building survey	Trevor Rushton
Employer's agent	Christopher Knott
Environment and specification	Tim Coffin
Fire Precautions (Workplace) Regulations 1997 and (Amendment) Regulations 1999	Ian McCammont
The Fire Precautions Act 1971	Ian McCammont
Floors in industrial and retail buildings	Graeme Lees
Fungi and timber infestation in the UK	Ian Ford
Glazing - windows and doors satisfying the Building Regulations	Trevor Rushton
Health and safety at work for surveyors	Paul Winstone, Philip Woods
High Alumina Cement concrete	Trevor Rushton
Identifying the age of buildings	Allen Gilham
Latent Damage Act 1986	Paul Lovelock
Life cycle costings	Kerr Houston, Mike Ridley
Mechanisms of water entry	Trevor Rushton
Non-destructive testing	Trevor Rushton
Partnering	Christopher Knott

Technical resources

Party wall procedure	Aidan Cosgrave, Paul Lovelock
Primary objectives of maintenance management	Mike Ridley
Problem areas with 1960s and 1970s buildings	Trevor Rushton
Procurement methods	Andrew Gear
Project management	Daniel Webb
Public Procurement, Public Private Partnerships and the Private Finance Initiative	Kerr Houston
Quality management and professional construction services	Aidan Cosgrave
The quantity surveyor's role	Stuart Russell
Radon	Tim Coffin
Reinstatement valuations for insurance purposes	Stuart Russell
Residential surveys	Trevor Rushton
Rights to light	Aidan Cosgrave, Paul Lovelock
Rising damp	Trevor Rushton
Rising groundwater	Trevor Rushton
Single project insurance	Jon Rowling
Site archaeology	Allen Gilham
Soil survey	Trevor Rushton
Soils and foundation design	Trevor Rushton
Sources of information in maintenance management	Angela Dawson
Specification writing	Tim French
Specifications	Tim French
Spontaneous glass fracturing	Trevor Rushton
Subsidence	Trevor Rushton
A systematic approach to maintenance management	Mike Ridley
Tender prices and building cost indices	Tim French
Troublesome plant growth: Japanese Knotweed	Keval Pankhania
Urban regeneration	Rob Sully
Useful information sources	Angela Dawson
Useful website addressses	Angela Dawson
VAT in the construction industry - zero rating	Stuart Russell
Vendor's surveys	Trevor Rushton
Watts and Partners' publications	Samantha Rumens
The Workplace (Health, Safety and Welfare) Regulations 1992 - Regulation 14	Trevor Rushton
Workplace Regulations - provision of sanitary facilities	Paul Winstone

Technical resources

Watts and Partners' publications

For further information on any of the following publications, please contact Watts and Partners' Marketing Department on +44 (0)20 7090 7820.

Watts Bulletin
As a monthly supplement to the Watts Pocket Handbook, the Watts Bulletin keeps its readership abreast of breaking news and changes in legislation within the property and construction industry. As part of the series, the practice also publishes two special issues a year, on topical subjects.

www.WattsHandbook.com
Watts Pocket Handbook owners are able to join the 'Watts Handbook Club'. Once registered at www.WattsHandbook.com, owners benefit from access to the archive of past issues of the Watts Bulletin and can subscribe to all future issues free by entering the special code on the Handbook bookmark. Furthermore, FAQs and technical queries can be answered by our expert in-house professionals. Visitors to the website can also find out more about the history of the Handbook, its technical contributors, and much more.

www.WattsandPartners.com
Visit www.WattsandPartners.com for more information on Watts and Partners and details of our range of services and our teams. The site is of interest to all stakeholders, from prospective employees to key clients, offering technical information, career opportunities, and case studies on a wide variety of jobs. A detailed search facility makes access to content quick and easy.

Watts Review
Watts Review 2004/2005 is due to be published in July 2005, subsequent to financial year-end. Published since 1999, this annual publication follows the format of an annual report and accounts document, updating clients on Watts and Partners' performance and development in the previous 12 months. The next issue will be available online and in hard copy.

Watts Practice Profile
The Practice Profile provides an account of Watts and Partners' offices, services, performance and procedures, and is presented as a matter of course when tendering for new business. As a control document for all other marketing publications, it is continually updated. The Practice Profile is tailored to the scope of our UK, Ireland and Iberian network of offices, with Irish and Spanish versions available.

Index

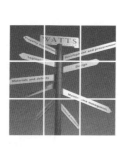

Index

TITLE	PAGE
1960s buildings	171, 175
1970s buildings	170
Above ground archaeology	12, 118
Access agreements	105-106
Acquisition	4, 8, 97, 98, 144, 238, 239, 241, 246, 247, 250, 251
Actionable injury	97
Acts of parliament and regulations	44-48
Adjudication	88-90, 92, 93
Adjudication under the Scheme for Contruction Contracts - how to get started	92-93
Air conditioning	81, 84, 136, 137, 217, 219
Air conditioning systems	136-137
Air handling units	138
Airtightness	58-59
Airtightness testing	59
Aluminium	165, 171, 178, 183, 192, 196, 200-202, 204, 205, 207
Appeals	70-73, 244
Appeals and called-in applications	70-71
Arbitration	44, 88, 90
Archaeology	12, 72, 116-118
Architectural and design criteria	109-120
Archival	11, 12
Asbestos	4, 20, 44, 46, 85, 154, 156-161, 171-173, 197
Basic design data	110-111
'Best value' in local authorities	229-230
Bi-metallic corrosion	165
Boilers	138, 217
Boreholes	122, 124, 125, 193
BREEAM	220-221, 256
British standard BS 6262	66
British standards	119, 122, 194, 232, 254
BUILD	6
Building Act	44, 45, 48
The Building (Amendment) Regulations 2004	44, 58
Building and construction regulations	57-67
Building control	10, 45
Building defects	85, 167-194
Building legislation and control	43-56
Building materials	191, 192, 197, 198, 215, 218
Building regulations	44, 46, 48, 55, 58, 161, 180, 200, 206, 217

Index

TITLE	PAGE
Building research	101, 182, 220, 254
The Building Research Environment Assessment Method	220
Building services	116, 218, 227, 230, 231, 237, 254
Building services design	135-141
Building survey	8, 10
Building types	111-114, 136, 170, 220
CAD	145, 148-150
Called-in applications	70
Capillarity	205, 206
Capital allowances	35, 44, 239-241, 251
Capital allowances for taxation purposes	239-241
Capital project advice	35
Carbonation	4, 163, 164, 173, 177, 179, 180, 184, 185, 196
CDM	5, 63-65
Cement	4, 155, 161-164, 173-175, 178, 196, 197, 254
CFCs	138, 219
Chemical and physical testing requirements	196-197
Chemical control	190
Chloride induced corrosion	163, 164, 184
Chloride Ion content	196, 197
Chlorides	163-164, 180, 181, 185, 191, 196
Circulars	72, 73
Cladding	9, 58, 133, 154, 165, 179, 180, 183, 199-209, 254
Commercial and industrial property	3-6
Commercial buildings	217, 224
Commercial/industrial surveys	4-6
Common defects in commercial and residential properties	168-169
Components	84, 132, 133, 141, 149, 157, 161, 165, 175, 180, 183, 204, 206, 224, 225, 232, 241, 250
Composite panels	4, 200-201
Computer-aided design	148-150
Concrete	103, 122, 131-133, 155, 156, 161-164, 168, 171-185, 196-198, 202, 205, 209, 216, 232, 238, 254, 257
Condition surveys	148, 225, 227-229
Conservation	4, 11, 46, 48, 58, 70-75, 117, 118, 211-221, 228, 236, 256

Index

TITLE	PAGE
Construction (Health, Safety and Welfare) Regulations 1996	44, 61-63
Construction contracts	29-31, 46-48, 88-92
The Construction (Design and Management) Regulations 1994	63-65
Construction management	29, 30, 38
Construction methods	131
Construction noise and vibration	106-107
Construction projects	29, 118
Contaminated land	5, 20, 44, 47, 157, 212-215, 242, 256
Contamination	9, 14, 19, 123, 125, 126, 138, 154, 164, 171, 191, 196, 212-214, 235, 242
Contract administrator	31, 219
The contract administrator's role	31-32
Contract management	33-36
Contracts and procurement	21-32
Contributors to the Watts Pocket Handbook	260-263
Control of asbestos at work	44, 46, 157
Control systems	139
Conversion formulae	152
Coordinating project information	118-120
Corrosion	155, 156, 162-165, 169, 171-178, 180, 182, 184, 191, 204, 205, 257
Corrosion of metals	165, 191
Cost management	14, 35, 233-252
Curtain walling systems	183, 201-205
Damage insurance	6
Damp	19, 132, 155, 156, 164, 168, 169, 174, 186-193
Data installations	141
Daylight and sunlight	101, 148
Daylight and sunlight amenity	101-102
Defects in concrete	184-185
Deleterious materials	8, 10, 154-156
Design	5, 6, 13, 14, 22, 25, 26, 29-32, 34-36, 38, 40, 44, 46, 47, 55, 58, 63, 64, 85, 89, 101, 104, 109-152, 155, 161, 162, 175, 194, 204-207, 213, 218, 220, 227, 235, 245-248, 252, 254, 256, 257
Design and build procurement	29, 34
Design data	110
Design loadings for buildings	126-130
The design process	36, 38, 64, 122, 220
Design specification	115

Index

TITLE	PAGE
Desktop studies	117
Development	6, 12-15, 23, 25, 26, 29, 34, 37-41, 45, 46, 70-72, 75, 97-101, 106, 116-118, 120, 123, 125, 131, 144, 148, 149, 170, 212-214, 221, 235, 239, 240, 245-247, 255
Development and procurement	21-41
Development applications and fees	70
Development monitoring	12-15
Development plans	70-72
Development plans and monitoring	70-71
Different specifications for different contracts	115
Dilapidations	9, 78-85
Diminution in value	80
Dioecious	189
Disability Discrimination	5, 8, 45, 47, 51-56, 80, 81
The Disability Discrimination Act 1995	45, 47, 51-52, 54
Discovery of building defects - statutory time limits	85-86
Dispute resolution	88, 92
Due diligence	7-10, 212, 241
Due diligence and the building survey	8-10
Employer's agent	34
Energy conservation	4, 46, 217-218, 228
Environment and specification	218-219
Environmental impact assessments	75
Environmental Protection Act	45, 107, 190, 212, 214
Equilibrium moisture content	192
Estate regeneration	234
Evidence in court	87
Expert witness	78, 86-88
Field evaluation	117
Finance monitoring	14
Finish	131, 132, 177, 179, 182, 192, 204
Fire Precautions	5, 45, 47-49, 112
The Fire Precautions Act 1971	45, 47-49, 112
Fire Precautions (Workplace) Regulations	45, 47, 49
The Fire Precautions (Workplace) Regulations 1997 and (Amendment) Regulations 1999	49-50
Fire safety	45, 50, 58, 200
Flatness	132
Floor slab	131-133, 174, 185, 202
Floors in industrial and retail buildings	130-133
Foundation	122, 132, 194, 213

Index

TITLE	PAGE
Funding	24-26, 28, 235, 251
Fungi and timber infestation in the UK	185-189
Gasket design	206
General Development Order and Use Classes Order	71-72
Geotechnical	122-125, 132, 213
Glass fracturing	208
Glazing	5, 8, 65, 66, 169, 171, 178, 202-204, 206-209, 218, 254
Glazing - windows and doors satisfying the building regulations	206-208
Gravity	205, 206
Groundwater	122-124, 126, 191-193, 212, 213
HAC	155, 161, 162, 196
HCFCs	138, 219
Health and safety	5, 6, 15, 20, 45, 50, 59-65, 160, 171, 177, 196, 197, 224, 228, 231, 254, 255
Health and safety at work for surveyors	59-61
High Alumina Cement concrete	161-163, 174
Historic building	11, 74
Housing Act	45, 48
Housing and residential property	17-20
Identification of radon	215
Identifying the age of buildings	11-12
Independent expert	78, 88
Industrial buildings	59, 130, 131, 172, 173, 240
Infestation	20, 185, 188, 189
Insect	20, 168, 188, 197
In-situ testing	124
Installations	6, 10, 141, 217, 219, 230, 240, 247
Joints	58, 131-133, 154-156, 172, 175-177, 183, 192, 201, 202, 205, 206, 216
Kinetic energy	205
Knotweed	9, 189, 190, 191
Land remediation allowances	242
Landlord and tenant	45, 80, 82, 83, 225, 252
Landscaping	190, 218, 227, 238
Latent Damage Act 1986	45, 85-86
Latent defects insurance	6
Legal and lease	77-93
Legal issues	9, 119

Index

TITLE	PAGE
Legislation	43-107, 116, 119, 157, 188, 214, 218, 229, 239, 240, 257
Life cycle costings	246-251
Lift terminology	139-140
Lighting	5, 8, 50, 62, 98, 100, 110, 140, 217, 238
Lighting design	140
Lighting requirements	110
Listed building	70, 71, 73, 74, 118, 236, 237
Listed building consent	70, 74, 118, 236, 237
Listed buildings and conservation areas	46, 73-75
Litigation	78, 88, 240
Loadings	4, 122, 126, 131
Local authority	11, 24, 30, 71, 75, 99, 100, 107, 112, 113, 213, 214, 252
Maintenance management	223-232
Materials	4, 8, 10, 12, 20, 40, 65, 66, 74, 89, 116, 119, 130, 131, 132, 153-165, 170, 179, 191, 192, 197, 198, 200-202, 208, 215, 217-219, 224, 226, 238, 245
Materials and components	224
Materials and defects	154-209
Mezzanine	25, 84, 131
Measured surveys	144-145
Mechanical control	190
Mechanisms of water entry	205-206
Mediation	88
Membranes	132
Monitoring service	13
Moulds	164, 171, 185-188
National Disability Council	51
NDT	197-198
Neighbourly matters	14, 95-107
Noise	19, 72, 105-107, 110, 137, 138
Noise and acoustics	110
Non-destructive testing	197-198
Nuisance	96, 106, 107
ODPM	70, 71, 73-75, 105, 234, 235
Offices	23, 46, 48, 49, 60, 61, 66, 110, 112, 127, 129, 136, 139, 140, 220, 238, 256

Index

TITLE	PAGE
Open book	36, 246
Partnering	30, 31, 35-36, 119
Party wall procedure	102-105
Payment	15, 22, 26, 28, 31, 32, 45, 48, 81, 89-91, 98, 106, 242
Performance related reward	36
Performance specification	115
PFI	22-28, 247, 257, 258
Planning policy	72
Planning policy guidance notes and circulars	72-73
Plant and equipment	138
Plaster/mortar	197
Polycarbonates	66
Polyurethane	200
Power systems	139
PPP	22, 28, 247, 258
Predicting costs	245
Preservation	74, 75
Pressure differential	206
Primary objectives of maintenance management	224
Problem areas with 1960s and 1970s buildings	170-183
Procurement and standard form contracts	29-31
Procurement methods	29, 38
Procurement process	24, 27, 35, 36, 246
Procurement strategy	226, 235
Project information	118-120
Project management	14, 30, 39-40, 119, 235, 254
Property acquisition	3-20, 144
The Provisions of Part II of the Housing Grants, Construction and Regeneration Act 1996	89-92
Public Procurement, Public Private Partnerships and the Private Finance Initiative	22-28
Public transport	51, 54
Quality management and professional construction services	40-41
The quantity surveyor's role	35
Radon	9, 20, 215-217, 258, 259
Rating revaluation 2005	244-245
Regeneration	45, 88, 89, 92, 234-235
Reinstatement valuations for insurance purposes	252
Repairs and defects	8
Residential properties	168, 215
Residential surveys	18-20

Index

TITLE	PAGE
Right of way	106
Rights to light	9, 96-100, 104, 105
Rising damp	132, 191-193
Rising groundwater	193
Sandwich panels	200, 201
Sanitary	5, 8, 19, 60, 62, 66
Schedule of Dilapidations	78-80
Scheduled ancient monument	11, 118
Section 18(1) of the Landlord and Tenant Act 1927	80, 83-85
Shops	46, 48, 60, 66, 67, 112
Single project insurance	6
Site analysis	143-145
Site archaeology	116
Soil survey	123-126
Soils and foundation design	122-123
Sources of information in maintenance management	230-232
Specification writing	115-116
Specifications	79, 115, 116, 119, 206
Spontaneous glass fracturing	208-209
Statutory time limits	85
Steel	89, 122, 124, 132, 155, 162, 163, 165, 169, 171-173, 179-184, 200, 202, 204, 209
Stick system	201
Structural and civil engineering design	121-133
Subsidence	19, 44, 172, 194
Sulphate attack	162, 178, 185
Sulphates in concrete	196
Surface tension	192, 205, 206
Surveys	4, 10, 18, 123, 132, 144, 148, 159, 188, 198, 225, 227-229, 259
Suspended	84, 131, 141, 201, 216
A systematic approach to maintenance management	224-227
Tables and statistics	151, 152
Technical resources	253-264
Tender prices	106, 243
Tender prices and building cost indices	243
Testing	4, 59, 66, 115, 122, 124-126, 132, 158, 161, 162, 195-198, 204, 205, 209, 238, 256, 259
Testing requirements	196
Thermal break	202, 206

Index

TITLE	PAGE
Thermal insulation	172, 175, 182, 217, 218
Thermography	198
Timber	11, 122, 168, 169, 177-180, 185-189, 191, 192, 197, 198, 207, 216, 218, 255
Town and country planning in England	69-75
Transaction structure	24, 27
Troublesome plant growth: Japanese Knotweed	189-191
Two stage tendering	36
Urban centre renewal	234
Urban regeneration	234-235
Use classes order	70, 71, 73
Useful information sources	254-255
Useful website addresses	255-259
Valuation	78, 80, 84, 99, 227, 244, 245, 252
Value for money	22, 23, 26, 28, 34, 35, 224, 227, 246, 247, 250
Variable air volume	136
VAT in the construction industry - zero rating	235-239
VAV	136, 137
Vendor's surveys	10
Warehouses	114, 130, 173, 175
Water chillers	138
Water entry	205
Watts and Partners' publications	264
Wildlife and Countryside Act, 1981	188, 190
Workplace (Health, safety and welfare) Regulations	46, 60, 65-66
The Workplace (Health, Safety and Welfare) Regulations 1992 - Regulation 14	65-66
Workplaces	49, 60, 67
Workplace Regulations - provision of sanitary facilities	66-67
Zero rating	235, 236

Notes

Notes

Notes

Watts and Partners

Consultants to the property and construction industry

Belfast
2-12 Montgomery Street
Belfast BT1 4NX
Tel: +44 (0)28 9024 8222
Fax: +44 (0)28 9024 8007

Bristol
33-35 Queen Square
Bristol BS1 4LU
Tel: +44 (0)117 927 5800
Fax: +44 (0)117 927 5810

Dublin
74 Fitzwilliam Lane
Dublin 2
Tel: +353 (0)1 703 8750
Fax: +353 (0)1 703 8751

Edinburgh
County House
63-65 Shandwick Place
Edinburgh EH2 4SD
Tel: +44 (0)131 229 9340
Fax: +44 (0)131 229 9800

Glasgow
176 Bath Street
Glasgow G2 4SE
Tel: +44 (0)141 353 2211
Fax: +44 (0)141 353 2277

Leeds
Atlas House
31 King Street
Leeds LS1 2HL
Tel: +44 (0)113 245 3555
Fax: +44 (0)113 245 1333

London (City)
1 Great Tower Street
London EC3R 5AA
Tel: +44 (0)20 7090 7820
Fax: +44 (0)20 7623 4450

London (West End)
11 Haymarket
London SW1Y 4BP
Tel: +44 (0)20 7533 4800
Fax: +44 (0)20 7839 4740

Madrid
Santa Engracia nº 151
planta 6ª
28003 MADRID
Tel: +34 91 4355459
Fax: +34 91 5758267

Manchester
60 Fountain Street
Manchester M2 2FE
Tel: +44 (0)161 831 6180
Fax: +44 (0)161 834 7750

And in association with
Turnbull Associés
Paris
Turnbull Associés
11, rue la Boétie
75008 Paris
Tel: +33 (0)1 47 42 31 38
Fax: +33 (0)1 40 06 04 87

www.WattsandPartners.com
www.WattsHandbook.com